4th Young Researcher Meeting – Trieste 2013

Journal of Physics: Conference Series Volume 470

Trieste, Italy
3-4 June 2013

ISBN: 978-1-62993-448-8
ISSN: 1742-6588

Printed from e-media with permission by:

Curran Associates, Inc.
57 Morehouse Lane
Red Hook, NY 12571

CURRAN ASSOCIATES INC. proceedings.com

Some format issues inherent in the e-media version may also appear in this print version.

Printed by Curran Associates, Inc. (2013)

Institute of Physics
Dirac House, Temple Back
Bristol BS1 6BE UK

Phone: 44 1 17 929 7481
Fax: 44 1 17 920 0979

techtracking@iop.org

Additional copies of this publication are available from:

Curran Associates, Inc.
57 Morehouse Lane
Red Hook, NY 12571 USA
Phone: 845-758-0400
Fax: 845-758-2634
Email: curran@proceedings.com
Web: www.proceedings.com

TABLE OF CONTENTS

4th Young Researcher Meeting, Trieste 2013 IOP Publishing
Journal of Physics: Conference Series **470** (2013) 012001 doi:10.1088/1742-6596/470/1/012001

Localization of inclusions in multiple prompt gamma ray analysis: a feasibility study

A Miceli[1], G Festa[2], R Senesi[1,3], G Gorini[2] and C Andreani[1]

[1] Department of Physics and NAST Centre, University of Rome Tor Vergata, via della Ricerca Scientifica 1, Rome 00133, Italy

[2] Department of Physics, University of Milan Bicocca, Piazza della Scienza 3, Milan 20126, Italy

[3] CNR-IPCF, Italy

E-mail: alice.miceli@uniroma2.it

Abstract. We investigate the feasibility of using low energy gamma rays from neutron capture to localize slabs inside samples. A new system based on two gamma ray detectors with 2D collimators to be tested at the INES beamline at the pulsed neutron source ISIS (Oxford, UK) is described. The system provides a localization of slabs inside extended samples by using gamma ray self-absorption. Geant4 Monte Carlo simulations of the beamline were carried out to model gamma spectra from test samples.

1. Introduction

Neutron capture prompt gamma activation analysis (PGAA) is a non-destructive analytical technique that provides information on the isotopic composition in samples. It is based on the measurements of gamma rays following neutron capture. The pattern of gamma rays, ranging from hundred keV to 12 MeV, is unique for each isotope [1]. The PGAA method has been applied in materials science, chemistry, geology, mining, archaeology, environment, food analysis, and medicine [2]. A current limitation of the technique is gamma ray self-absorption, causing attenuation of gamma rays within the sample itself. That may lead to significant errors [1]. In this paper, we exploit the gamma ray self-absorption and investigate the feasibility of using low energy gamma rays (< 650 keV) to localize metal slabs inside extended objects. Depending on the position of the inclusion, the gamma rays will be attenuated differently. Using two back-to-back detectors on opposite sides of the sample we can measure the ratio of the gamma ray intensities and calculate the approximate position of the slab. Spatial distribution of elements has recently been measured by scanning a sample with a collimated neutron beam and by using a collimated gamma detector [3]. Drawback of the technique is the high collimation of source and detector needed and the long beamtime required. The method described in this paper will make use of the full neutron beam area and allow fast localization of the slabs along the direction where the gamma ray detectors are placed. Information on the position of the inclusion could be used to correct for the self-absorption and obtain more precise quantitative analysis. This study has been performed with Monte Carlo simulations based on the GEANT4 code [4, 5].

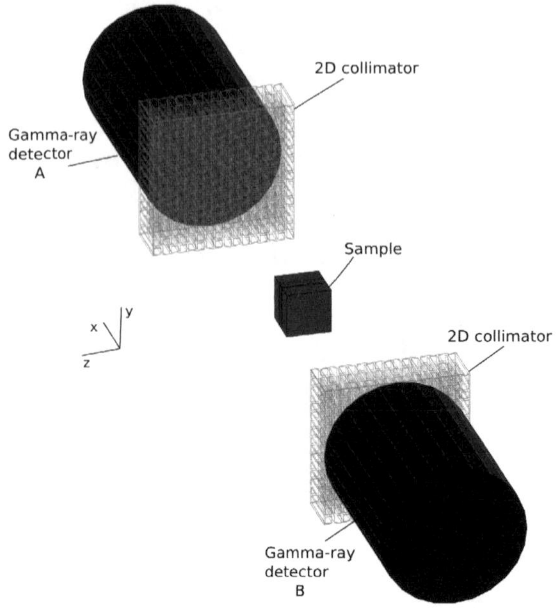

Figure 1. Simulated geometry. Gamma ray detectors, 2D lead collimators, and sample are illustrated. The neutron beam was along the z axis. The sample consisted of a matrix (Fe or Al) with one or two slabs (Cu, Fe). Sample thickness in the x direction was D.

2. Materials and Methods

2.1. Theory of measurements

The attenuation of gamma rays within the sample increases with thickness, density, and atomic number of attenuating medium, and with the decreasing of the gamma ray energy [1]. For gamma rays penetrating a uniform layer of material in the normal direction, the transmitted flux is:

$$I = I_0 \exp(-\mu x) \tag{1}$$

where I_0 is the incident neutron flux, μ is the attenuation coefficient, and x is the layer thickness. In a neutron capture event the gamma rays are emitted isotropically. In order to ensure that the gamma ray detectors measure intensities of gamma rays reaching the detectors after traveling along a straight-line path from the gamma ray source, i.e. the inclusion, we employed 2D collimators. The collimators were located immediately in front of the detectors. The ratio of counts on the gamma ray detectors A and B can therefore be expressed as:

$$\begin{aligned} \frac{C_A}{C_B}(E) = \frac{I_0(E) \times \epsilon_A(E) \times \exp(-\sum_{i=1}^{N} \mu_i(E) \times x_i^A)}{I_0(E) \times \epsilon_B(E) \times \exp(-\sum_{i=1}^{N} \mu_i(E) \times x_i^B)} = \\ = \frac{\epsilon_A}{\epsilon_B}(E) \exp\left[-\sum_{i=1}^{N} \mu_i(E) \times (x_i^A - x_i^B)\right] \end{aligned} \tag{2}$$

where $I_0(E)$ is the number of the gamma rays emitted with energy E, $\epsilon_A(E)$ is the efficiency of the detector A, $\epsilon_B(E)$ is the efficiency of the detector B, μ_i is the attenuation coefficient of the material i, x_i^A is the distance traveled by the gamma ray in the material i in the direction of the detector A, and x_i^B in the direction of the detector B.

For samples in which the matrix is uniform and the slabs are small enough to ignore absorption within the slabs themselves, we can calculate the position of the slabs, knowing that $x_i^A + x_i^B = D$, as:

$$x^A = \frac{1}{2} \times \left[D - \frac{1}{\mu_M(E)} \ln \left(\frac{C_A(E) \times \epsilon_B(E)}{C_B(E) \times \epsilon_A(E)} \right) \right] \tag{3}$$

where μ_M is the attenuation coefficient of the matrix, and x^A is the distance of the center of mass of the slab from the sample edge along the x axis toward the detector A, D is the sample thickness in the direction x. The position of the slab will be:

$$x^{slab} = D/2 - x^A = \frac{1}{2 \times \mu_M(E)} \ln \left(\frac{C_A(E) \times \epsilon_B(E)}{C_B(E) \times \epsilon_A(E)} \right) \tag{4}$$

The values of $x^{slab}$are independent of the gamma ray energy. For that reason, to decrease uncertainties we averaged on all the values of x^{slab}.

This procedure was applied to several samples of Fe and Al with slabs of thickness 0.1 cm at different positions.

2.2. Monte Carlo simulation

The proposed multi prompt gamma ray analysis system consists of two back-to-back gamma ray detectors and two 2D collimators located immediately in front of the detectors. Detectors diameter and length were 6.4 cm and 9 cm, respectively. The gamma ray detectors were placed perpendicular to the primary beam at opposite sides of the sample at a distance of 15 cm. The 2D lead collimators were 2.2 cm thick with 0.1 cm septa and 0.4 cm periodicity. Figure 1 shows the simulated geometry. The simulation was carried out for a water moderator at 293 K at a spallation source. The neutron energy spectrum is shown in figure 2. To speed up the simulation we selected neutron energies up to epithermal range (< 1.2 keV). The neutron beam was simulated as a parallel beam of size 2 x 2 cm^2.

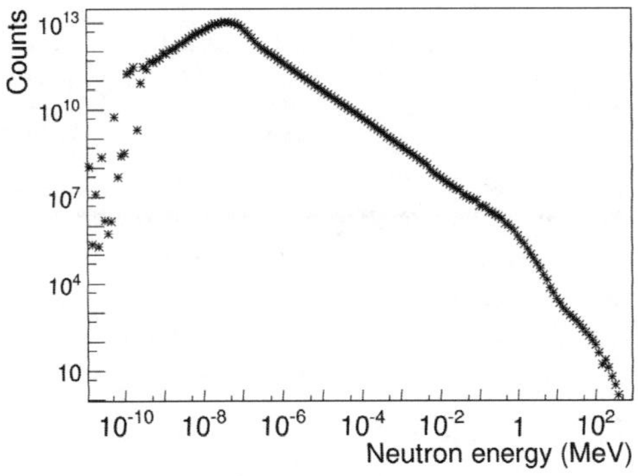

Figure 2. Neutron energy spectrum simulated with the Monte Carlo radiation transport code MCNPX (private communication with S. Ansell [6, 7]).

Samples with one and two slabs were simulated to show the capability of the system to localize slabs inside a matrix representing a model of a realistic sample. The matrix size was 2 x 2 x 2 cm^3. The matrix material was Fe or Al. The slab size was 0.1 x 2 x 2 cm^3. The slab material was Cu and/or Fe. Neutron tracks were followed through the simulated geometry until they were absorbed or exited the world volume. Inelastic and elastic scattering were simulated with the high precision neutron model of Geant4 [8]. The neutron capture was modeled by randomly

Journal of Physics: Conference Series **470** (2013) 012001 doi:10.1088/1742-6596/470/1/012001

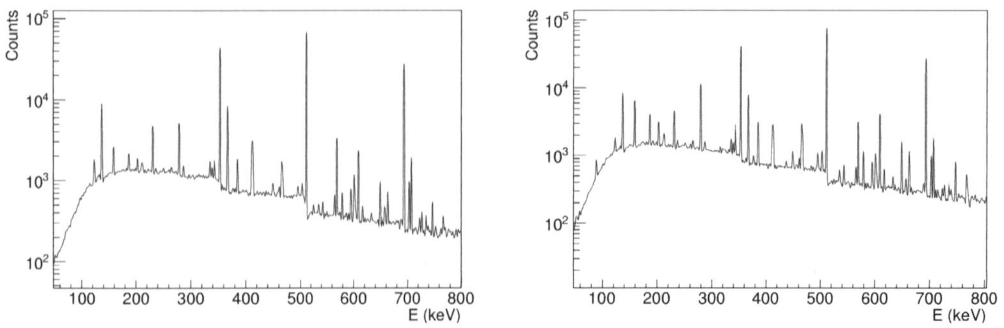

Figure 3. Simulated energy spectrum measured by detector A (left) and B (right) from a Fe sample with a Cu slab in the energy range [0, 800] keV.

sampling gamma ray energies from the nuclear decay schemes (Evaluated Nuclear Structure Data File [9]). The radioactive decay of unstable nuclei produced by neutron capture was simulated with the Geant4 radioactive model [8]. Electromagnetic interactions were simulated with the Geant4 Livermore model [8]. Energy and measurement time were recorded for every gamma ray detected. To evaluate the feasibility of the method without loss of generality we considered $\epsilon_A(E) = \epsilon_B(E) = 1$ in equation 4. The number of simulated events was 10^9.

3. Results

Figure 3 shows a simulated energy spectrum measured by detector A and B from a Fe sample with a Cu slab. Gamma ray lines from neutron capture in Fe and Cu are clearly visible in the energy range [0, 800] keV.

Figure 4 shows the ratio of intensities of Cu lines as a function of gamma ray energies for slabs in a Fe matrix placed at offset -0.2 cm and 0.2 cm along the x axis. The ratio C_A/C_B is well separated for the two offsets for all the gamma ray lines.

Table 1 lists the reconstructed position of the slabs calculated from equation 4 for different samples. The error on the calculated Cu slab position is the standard deviation of the x^{slab} position calculated for each peak. The Fe element shows only a gamma-ray peak, at 352 keV, in the selected energy range (< 650 keV) in accordance with the prompt gamma-ray spectrum catalog [1] in which only the 352 keV peak has partial gamma ray production cross section above 0.1%. The difference between the actual and reconstructed position of the slabs is ≤ 0.1 cm for all the studied configurations. Samples containing less abundant and/or weaker gamma-emitters could be investigated using the neutron time of flight (TOF) information and the resonance structure of neutron capture cross section. Gamma ray lines of less abundant or weaker gamma-emitter isotopes could be discerned by selecting gamma ray spectra at specific TOF windows. In future, simulations accounting for detector effects and gamma-ray background will be performed in order to select the most suitable detector and to study limits and geometrical uncertainties of the technique for different matrix/slab materials and thicknesses. Measurements at the pulsed neutron source ISIS (Oxford, UK) are foreseen.

4. Conclusions

The feasibility of using low energy gamma rays from neutron capture to localize metallic slabs inside a sample has been investigated with Monte Carlo simulations. A detection system consisting of two back-to-back gamma ray detectors with two 2D collimators has been proposed.

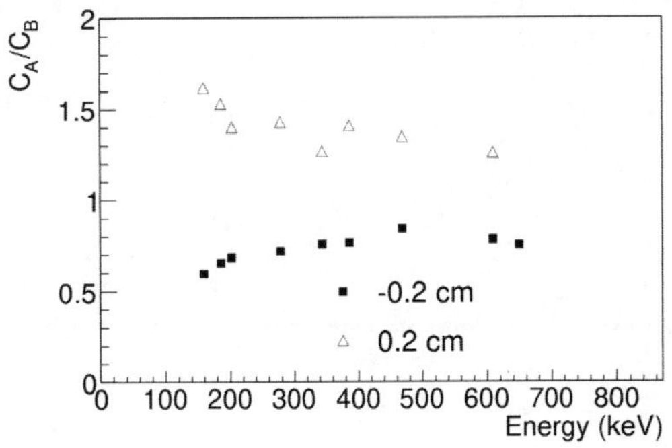

Figure 4. Ratio of counts on detector A and detector B as a function of gamma ray energies for a Cu slab in a Fe matrix placed at -0.2 cm (black full square) and 0.2 cm (red empty triangle) offset along x.

Table 1. Calculated slab positions x_R^{slab1} and x_R^{slab2}.

sample number	matrix material	slab1 material	slab2 material	x^{slab1} (cm)	x^{slab2} (cm)	x_R^{slab1} (cm)	x_R^{slab2} (cm)
1	Fe	Cu	-	-0.5	-	-0.46 ± 0.05	-
2	Fe	Cu	-	$+0.5$	-	$+0.46 \pm 0.07$	-
3	Fe	Cu	-	0	-	-0.03 ± 0.02	-
4	Fe	Cu	-	$+0.2$	-	$+0.21 \pm 0.02$	-
5	Fe	Cu	-	-0.2	-	-0.19 ± 0.03	-
6	Al	Cu	Fe	$+0.2$	-0.2	$+0.2 \pm 0.1$	-0.3
7	Al	Cu	Fe	$+0.4$	-0.4	$+0.5 \pm 0.3$	-0.3

The position of the slab has been correctly reconstructed using the ratio of the intensity of two gamma ray detectors placed on opposite sides of the sample. Simulations taking into account detector effects and studying the limitations of the techniques are ongoing. Information on inclusion position could be used to correct for the gamma ray self-absorption and to obtain more precise quantitative analysis. Possible applications of the technique are in metallography, geology, and archeology.

5. Acknowledgments
Computing resources of the Science and Technology Facilities Council's e-Science facility were used in this research.

References
[1] Molnar G L 2004 *Handbook of Prompt Gamma Activation Analysis with neutron beams* (Kluwer Academic Publishers) ISBN 1-4020-1304-3
[2] Perry L A, Firestone R B, Molnar G L, Revay Z, Kasztovszky Z, Gatti R C and Wilde P 2002 *J. Anal. At. Spectrom.* **17** 32–37
[3] Belgya T, Szentmiklosi L, Kis Z, Kasztovszky Z, Schulze R, Kudejova P, Materna T, Canella L and Jolie J 2009 *Proceedings of the 14th International Conference on Cultural Heritage and New Technologies*
[4] Agostinelli S *et al.* 2003 *Nucl. Instr. and Meth. A* **506** 250 – 303
[5] Allison J *et al.* 2006 *Nuclear Science, IEEE Transactions on* **53** 270–278

[6] Bedogni R, Esposito A, Andreani C, Senesi R, Pascale M P D, Picozza P, Pietropaolo A, Gorini G, Frost C D and Ansell S 2009 *Nucl. Instr. and Meth. A* **612** 143 – 148
[7] Rebai M, Gorini G, Perelli-Cippo E, Pietropaolo A, Tardocchi M, Fazzi A, Milani E, Verona-Rinati G, Andreani C, Senesi R, Frost C D, Schoonevled E M, Rhodes N J, Bedogni R and Esposito A 2010 *XIX ICANS Conference Proceeding*
[8] Geant4 collaboration 2011 *Physics Reference Manual*
[9] Tuli J K 2001 *Evaluated Nuclear Structure Data File* (Upton NY 11973-5000 USA)

Journal of Physics: Conference Series **470** (2013) 012002 doi:10.1088/1742-6596/470/1/012002

Watch Dog detector for beam diagnostic in hadrontherapy application

E Rojatti[1,2], G Bazzano[2], J Bosser[2,3], M Caldara[4], G Calvi[2], L Lanzavecchia[2], A Parravicini[2], M Pullia[2] and C Viviani[2]

[1] Dipartimento di Fisica, *Università degli studi di Pavia* - Pavia,Italy
[2] Centro Nazionale di Adroterapia Oncologica (CNAO) - Pavia, Italy
[3] CERN - Switzerland
[4] Dipartimento di Ingegneria, *Università degli studi di Bergamo* - Bergamo, Italy

E-mail: elisa.rojatti01@ateneopv.it

Abstract. The "Watch Dog" is a beam monitor designed for medical accelerators, that will be installed at the end of the CNAO (Centro Nazionale di Adroterapia Oncologica) extraction lines. Its main goal is to achieve a real time monitoring of the beam position during patient treatments; the system can generate an interlock signal in case the measured quantity is out of the nominal range. In this paper the Watch Dog is described, and preliminary tests are presented.

1. Introduction

Hadrontherapy [1] is a radiotherapy technique which uses non elementary charged particles (in particular protons and carbon ions) instead of photons or electrons.
The CNAO (Centro Nazionale di Adroterapia Oncologica) is the first Italian Centre for deep hadrontherapy. Patient treatments began in September 2011 with protons and in November 2012 with carbon ions. An elaborated beam control is required to ensure patient safety and a correct dose delivery to the tumor volume. For this purpose a sizeable amount of beam diagnostic instrumentation has been developed [2].
The "Watch Dog" (WD) is a position and profile beam monitor, based on the scintillation process, that will be installed at the end of the CNAO extraction lines. The hardware is designed, built and assembled and a first release of motion and data acquisition software is ready. Preliminary laboratory tests and with beam have been performed.
The first part of the paper presents the CNAO facility and the HEBT (High Energy Beam Transfer) lines. The second one describes the WD detector. The third one deals with preliminary tests in lab and with beam. In the conclusions the open points are discussed.

2. The CNAO facility and the HEBT lines

The CNAO accelerator (figure 1) consists mainly of a 25 m diameter synchrotron that accelerates the proton or the carbon ion beam until the desired kinetic energy. Four HEBT (High Energy Beam Transfer) lines (three horizontal and one vertical) (figure 2), deliver the accelerated beam to three treatment rooms. The beam is continuously extracted from the synchrotron with a slow extraction process, during a period variable from 1 to 10 seconds. The range of extraction

energies for protons is 60 to 250 MeV, while that for carbon ions is 120 to 400 MeV/u.
The dose is delivered to the patient using an active technique. The target volume is divided in slices, each one corresponding to a defined beam kinetic energy, and the pencil beam is deflected by two fast magnets on the horizontal and the vertical plane in order to "paint" each slice. Beam size at the entrance of the two Scanning Magnets varies with the energy of the particles from 4 to 10 mm for each direction (horizontal and vertical) independently. Beam position at the entrance shall be fixed with a great precision (between 0.1 and 0.2 mm) to allow the Dose Delivery System ([3], [4]) to work properly and efficiently[1].

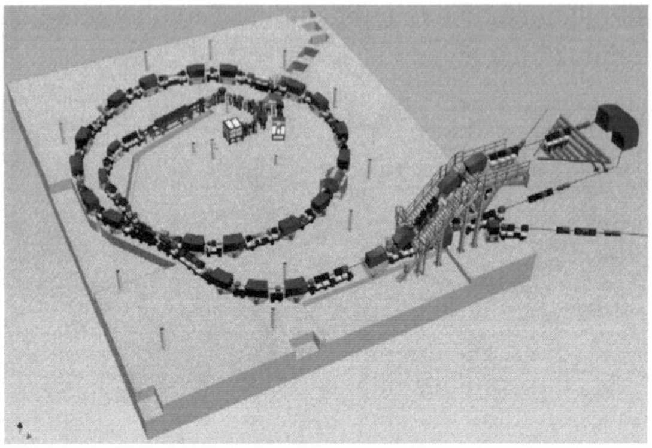

Figure 1: The CNAO facility.

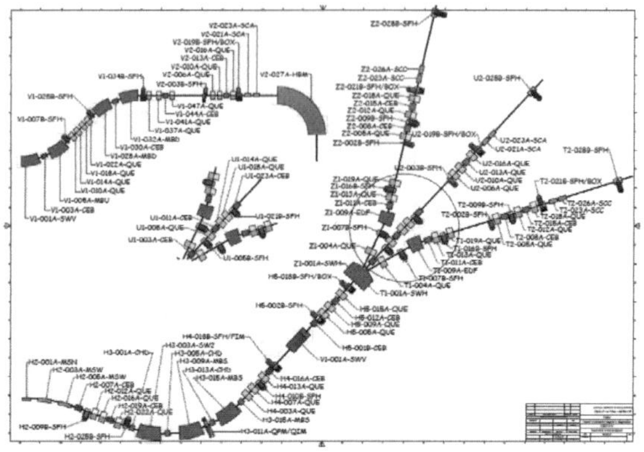

Figure 2: The HEBT lines.

3. The Watch Dog detector

The WD detector (figure 3) is a beam monitor that will be installed almost at the end of each HEBT line, just before the Scanning Magnets entrance (before the thin window, i.e. under vacuum). Its main goal is to monitor in real time and in a non destructive way the beam position and to generate an interlock signal if the measured quantity is out of the nominal range. It may also scan the full beam spot to provide horizontal and vertical beam profiles.
The WD consists of two couples of scintillating fibers (type SCSF-3HF(1500)) displaced transversely to the beam direction, in the horizontal and the vertical beam planes, and coupled to Avalanche Photo-Diodes (APDs, made by Hamamatsu, model S8664-10K). Particles crossing fibers release energy and produce light which is converted by the APDs to a current signal proportional to the number of particles crossing the fiber.
Each fiber is mounted on a movable mechanism, actuated by a brushless motor. Fiber position measurement is constantly provided by a linear potentiometer, whose analog signal output is proportional to the absolute fiber position.

3.1. Use: Watch Dog Mode and Scanning Mode

This detector can be used both in Watch Dog and in Scanning Mode. In the first case the WD can be used during patient treatments, because it doesn't perturb the beam. Fibers are moved up to a fixed position, at beam spot borders, in order to intercept only the tails of the beam. The fiber positions are decided *a priori* according to the beam characteristics and the information about the position is distributed by the Timing System before each cycle. The beam center of

[1] The dose must be delivered to the target with a 2% precision. The Dose Delivery System is designed in order to give an interlock if the position of the beam on the target is out of the range that guarantees such a precision on the dose delivered.

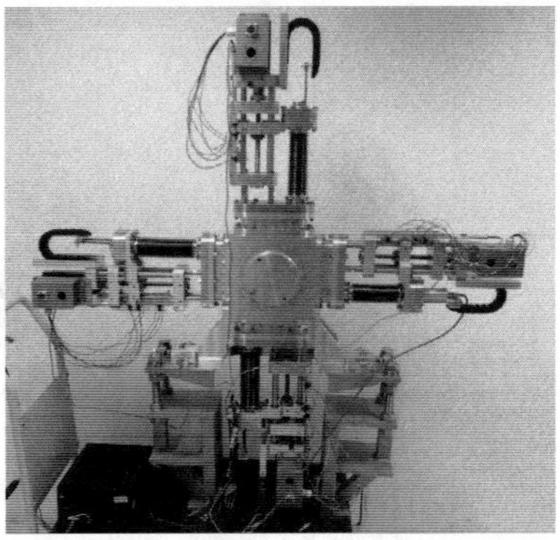

Figure 3: WD detector layout.

gravity (c.g.) can be reconstructed by measuring the normalized quantity:

$$\frac{I_1 - I_2}{I_1 + I_2} \propto c.g. \tag{1}$$

where I_1 and I_2 are the output signals of two APDs connected to a couple of parallel fibers. In the second case each fiber scans the entire beam spot. The transverse beam profile can be reconstructed by plotting $I_{fiber}\ Vs\ Position_{fiber}$. In this modality the WD cannot be used during patient treatments because it perturbs the beam.

4. Preliminary results with a light source

Preliminary tests on the WD were performed to study the detector response to a light source and its performances. A collimated light beam[2] was delivered in the WD vacuum chamber in place of the beam; the potentiometer output signal was acquired synchronously with the APD output signal. The first one is proportional to the fiber position, and the second one is converted to a voltage signal by a current-to-voltage amplifier. Figure 4 shows the light profiles obtained by scanning the light beam with the fiber, and plotted after the raw data processing. Each profile corresponds to a particular value of the High Voltage provided to the APD.

4.1. APD gain curve and dark current trend

Figure 5 shows the APD gain and the dark current trend as function of HV, expressed in decibel[3]. The gain was obtained by processing data used to reconstruct the light profiles. For each value of the HV provided (from 200 V to 440 V), the APD output signal was integrated over the number of samples and then scaled with respect to the integral value obtained for HV equal to 200 V. To perform dark current measurements the fiber was shielded from the ambient light and placed in a fixed position. Data were acquired by changing the HV provided to the APD from 200 V to 440 V. The dark current curve was obtained by integrating the APD output signal over the number of samples[4] for each HV value, and by scaling with respect to the integral value obtained for HV equal to 200 V [5]. We find that the experimental curves are in good agreement

[2] The light source was collimated through an alluminium plate with a hole (1 cm of diameter) placed in front of the tank (housing the fibers) aperture.
[3] With reference to 35.61 V, the integral value calculated for an APD supply voltage of 200V.
[4] 8000 samples were acquired with a 5kHz rate both for the gain and for the dark current measurements.
[5] The dark current value for the APD supplied with 200V is 60pA (from the APD datasheet).

Journal of Physics: Conference Series **470** (2013) 012002 doi:10.1088/1742-6596/470/1/012002

with those reported on the APD data sheet.

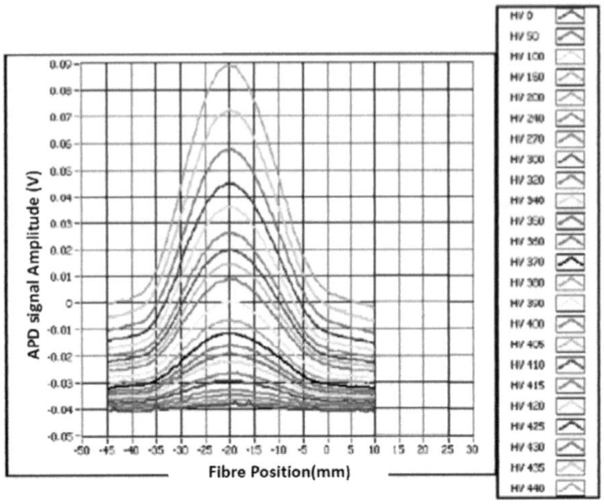

Figure 4: Light Profiles obtained by scanning the light beam with one fiber (plot obtained after raw data processing). Each colored line corresponds to a different value of the High Voltage provided to the APD.

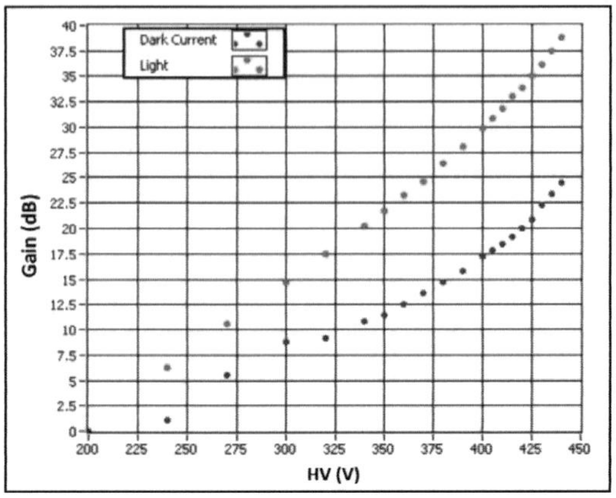

Figure 5: APD gain curve and dark current trend in the provided HV range 200 V to 440 V.

4.2. WD spatial resolution and accuracy

For this test the APD was supplied by 420 V, in order to obtain a quite high signal amplitude at the output. The light was sent to the detector sensitive area through a vertical slit. The slit was horizontally displaced in ten different equally spaced fixed positions. For each configuration ten light beam scans were performed, and the peak average position and the standard deviation were computed: they are reported on the vertical axis in figure 6. Standard deviation values are quite small (of the order of tenths of mm), which implies a good spatial resolution. On the horizontal axis of the plot the ten horizontal slit positions are reported, with the error due to the imprecision in displacing the slit ($\pm$ 1mm). We observe a linear relation between the slit position and the peak average position, that implies also a good accuracy of the detector in measuring a beam position displacement.

5. Preliminary beam tests

In beam tests the WD was placed in air, at the isocenter of a treatment room. One fixed fiber, placed at the center of the active area, was aligned with the proton beam. The APD was supplied by 420 V and the output signal was acquired. Asynchronous acquisitions were performed every six seconds[6] with the beam and they are compared with dark current measurements. Figure 7 shows data obtained by sampling at 5 KHz (raw data). By averaging the acquired samples over a 20 ms period it is possible to obtain a good signal to noise ratio, as shown in figure 8, where the spill duration (about one second) is also visible.

6. Conclusions

The usefulness and the innovation of the WD are related to its capability in monitoring on-line the beam position during patient treatment giving a fast interlock in the case it is out of the

[6] While beam cycles are of five seconds.

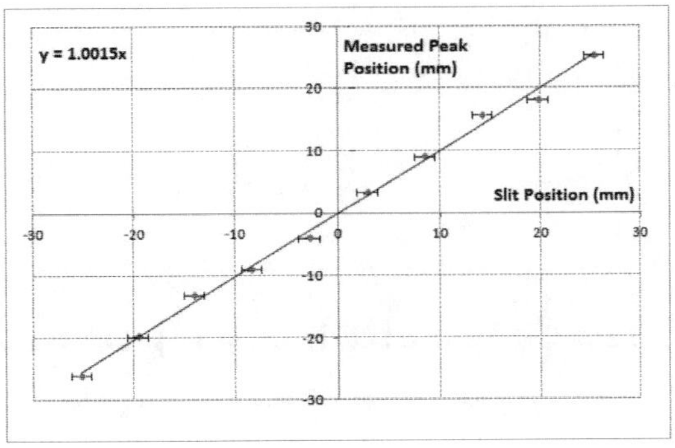

Figure 6: Measured peak position (mm) plotted versus the slit position (mm). A linear fit is performed on data. The equation of the fit line is expressed in red on the top.

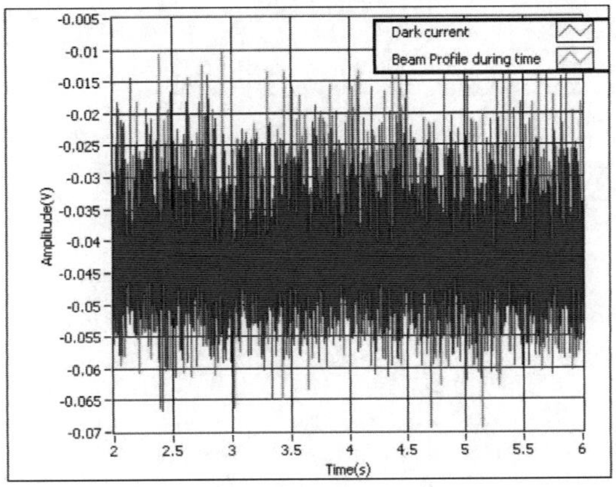

Figure 7: APD output amplitude during time for an acquisition rate of 5 kHz. The red line is the signal generated by the beam, the blue line is the dark current measurement.

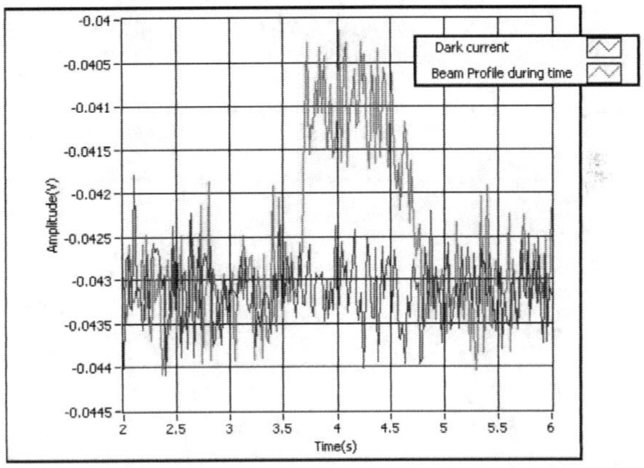

Figure 8: APD output amplitude during time after raw data processing: the acquired samples were averaged over a period of 20 ms. The red line corresponds to the signal generated by the beam, while the blue line is the dark current measurement

nominal range. Until now a prototype has been built and assembled and a first version of the motion and acquisition software has been delivered. Preliminary tests have been performed to study the detector response and its performances and to check the acquisition software. More beam tests will be performed in the next future (i.e. beam scans, tests with different particles and energies, time resolution of the detector and time response), the software will be integrated in the accelerator Control System, and the detector will be eventually installed on the beam line.

References

[1] Schardt D, Elssser T and Schulz-Ertner D 2010 Heavy-ion tumor therapy: Physical and radiobiological benefits *Rev. Mod. Phys.* **82** 383-425
[2] Caracciolo H et al 2011 Beam Diagnostic Commissioning at CNAO *Proc. of IPAC* (San Sebastian, Spain) 2848-2850.
[3] Giordanengo S et al 2008 The CNAO system to monitor and control hadron beams for therapy *Nuclear Science Symp. Conf. Record* (Dresden, Germany) 5609-5611.
[4] Giordanengo S et al 2011 Commissioning operations and performances of the Dose Delivery System for CNAO *Nuclear Science Symp. and Medical Imaging Conf.* (Valentia, Spain) 2811-2814.

Morphological investigation and physical characterization of ancient fragments of pyrogenic carbon

E Pusceddu[1], **I Criscuoli**[2] **and F Miglietta**[1,2]

[1]Istituto di Biometeorologia del Consiglio Nazionale delle Ricerche, IBIMET CNR,Via Giovanni Caproni, 8 - 50145 Firenze, Italy

[2]FoxLab, Fondazione E. Mach, via E. Mach 1, 38010 S. Michele all'Adige (TN), Italy

E-mail: e.pusceddu@ibimet.cnr.it

Abstract. In the latest years, the attention toward the use of pyrogenic carbon as a climate mitigation strategy has increasingly grown. Biochar (BC) contains substantial amount (60-90%) of pyrogenic carbon, which is a recalcitrant material and it is hardly decomposed by biotic and abiotic oxidation. The carbon mitigation potential of biochar is associated to the fact that carbon is not easily released back into the atmosphere, even after very long incubation time in the soil. Several studies have been addressing the understanding of the fate of pyrogenic carbon in the soil in a quantitative way, but only a few actually considered materials that were produced in the past and they were not fully able to estimate the fraction of carbon that was oxidized on centennial time scales. In this paper, an old deposits of biochar in soils of the Eastern Alps (Trentino, Val di Pejo) was dated at 1859 by means of a dendroanthracological approach. Carbon decomposition in those soils was then investigated to calculate the fraction of carbon that was lost over 155 years. Part of this study is focused on the morphological and physical characterization of several fragments of biochar, using a scanning electron microscope (SEM). Such study enabled the identification of specific morphological features of tracheids in the old biochar, which were tentatively associated to a differential oxidation of the structures that were created during carbonization from lignin and cellulose.

1. Introduction

In the last decades, the growing concern for global climate changes multiplied the efforts of the scientific community to understand the impact of those changes on flora, fauna and the humans and to identify potential adaptation and mitigation strategies. The cause of the global mean temperature rise and of the related augmented climate variability is attributed to the increasing anthropogenic Green House Gases (GHGs) presence in atmosphere. As reported by the Intergovernmental Panel on Climate Change in 2007 [1], the main effects of continued GHG emissions are: rising sea levels; damage to many global ecosystems; increase of droughts and heavy precipitation events; rising acid levels in the oceans. Biochar based strategies, which consider the application of large amounts of charred material to agricultural soils have been identified among those methods able to mitigate climate changes and in a recent study Woolf et al. (2010) [2] quantified the theoretical carbon sequestration potential of biochar following its incorporation into agricultural soils as large as 1.8 Gt C per year.

Journal of Physics: Conference Series **470** (2013) 012003 doi:10.1088/1742-6596/470/1/012003

Currently, a wide range of disciplines are involved in understanding the potential feedbacks of these strategies to improve biochar characterization, and to understand its properties as soil amendment in laboratory and field experiments [3, 4]. Biochar is produced by the thermo-chemical conversion of biomass under the conditions of low temperature (500°C ÷ 600°C) and little amount of oxygen (pyrolysis process). Different kinds of feedstocks can be used to produce biochar (agricultural residues, biomass crops and agroforestry products) in an ideal ecological-agro-climate cycle increasing soil fertility, sequestering organic carbon and producing energy via the pyrolysis exothermic reaction [5, 6]. Little information exists about the recalcitrance of biochar, and most of investigations were made on relatively young biochars (no more than 3 years old). Starting from the recent identification of a biochar kiln platform in Trentino, Val di Pejo (Eastern Alps in north of Italy) dated 1859 it was possible to extract ancient biochar and soil samples, and to perform an exhaustive characterization of the recalcitrance of an ancient and exactly dated biochar.

Our study aims at the physical and morphological characterization of biochar. In particular, we have compared the old biochar with the modern one, studying their density and morphological structure trying to understand if the biochar is really recalcitrant in soil on a long time scale.

2. Materials and methods

Samples of old biochar were taken in Val Pejo (Trentino, Eastern Alps, North Italy). They have been dated 1859 and their feedstock was coniferous larch wood. Using the same feedstock, a modern biochar was produced through a thermo-chemical process, which resembles the one that formed the old biochar: approximately 600°C in oxygen free atmosphere [7]. Before any investigation, both old and modern biochar underwent a thermal treatment for 48h at the temperature of about 80°C. The dry samples were cut in order to have a homogenous shape and each sample was weighed. The density data were collected analyzing 30 samples for each kind of biochar. This carbon based material has a high porosity, as analyzed by a penetrometer instrument under low pressure at about 10^{+4} Pa.

Morphology investigations were performed by scanning electron microscope (SEM) technique. We used XL 20 FEI SEM, with CRYO-GATAN ALTO 2100 technology. The measurements were performed on the samples dried under vacuum, following the common procedure. To study the inflation of the pores of old and modern biochar, the modern biochar was oxidized to simulate aging. We produced two samples using a H_2O_2 thermal bath at temperature of 30°C for 72h and at 75°C for 120h. The banding deformation of the pores was verified by SEM analysis. Several images were collected for each sample, in order to have accurate information concerning different areas on the internal surface of the biochar.

3. Results and discussion

SEM analysis revealed that biochar samples were the result of thermo-chemical transformation of larch wood. This conclusion is supported by a microstructure of paired lenticels[1] that is universally considered a typical discriminant for the identification of the wood of this conifer species [8]. In figure 1, we show the SEM image of a cross-section of old biochar, where the lenticels appear in pairs.

After drying, old and modern biochar were weighed and density data were collected by analyzing 30 samples for each old and modern biochar. This carbon based material is characterized as highly porous, as analyzed by penetrometer instrument under low pressure at about 10^{+4} Pa. We collected data of both the volume of carbon and the empty pore space volume. The averaged density of the old biochar $\langle\rho_{old-BC}\rangle$ is about twice the density of the

[1] A porous tissue consisting of cells with large intercellular spaces in the periderm of the secondarily thickened organs.

Figure 1. SEM image of transverse section of old biochar. The lenticels inside the pores appear coupled, as highlighted in the black circles.

modern biochar $\langle \rho_{mod-BC} \rangle$:

$$\langle \rho_{old-BC} \rangle = 0.42 \pm 0.08 \text{ gr/cm}^3,$$
$$\langle \rho_{mod-BC} \rangle = 0.24 \pm 0.03 \text{ gr/cm}^3.$$

We have considered the ratio (RT) between the pores thickness and the internal radius. We calculated that the average ratio for old biochar is almost twice that of the modern biochar:

$$RT_{old-BC} = 0.36 \pm 0.01,$$
$$RT_{mod-BC} = 0.21 \pm 0.01.$$

These results seem very peculiar because both biochars come from the same feedstock and the modern biochar has been produced trying to replicate as closely as possible the chemical process and the thermo-chemical conditions undergone by the old biochar. To study the different density between old and modern biochar a SEM investigation was performed. In figure 2, SEM images for both biochars are shown. The different porosities we measured could signify, in fact, that in old biochar the pores have thicker walls than the pores in modern biochar. These results have been achieved analyzing the SEM images in 4 different areas of each biochar sample. On these images, that were collected for 30 old and modern biochar samples, a statistical analysis has been performed. The pores appear thicker in old biochar and this is coherent to the observation of an increased density.

Biochar is composed of carbon aromatic rings, and Kinoshita et al. found that aromatic molecules, when oxidized, are modified due to bending of the plane structure [9]. Indeed, the oxidation of aromatic rings often translates into increased molar mass (m.m.), as for benzene (C_6H_6 m.m. = 78.11 gr/mol) oxidized to 1,2-dihydroxybenzene ($C_6H_6(OH)_2$, m.m. = 110,1 gr/mol). Is the observed inflation effect associated to oxidation of the old biochar in soil over long time scales?

In order to verify our hypothesis concerning the oxidation mechanism in old biochar, an oxidation process in modern biochar has been done. Before performing the chemical oxidation process, SEM investigation was repeated on modern larch biochar; afterwards, modern larch biochar samples were treated by H_2O_2 bath at room temperature for 72h and were then analyzed by SEM, as shown in figure 3 a) and b).

Comparing the results of the modern biochar with and without oxidation, we observe no significant difference. We presumed that the oxidation process, in terms of residence time and temperature, was not sufficiently strong in order to obtain the effects observed in the old biochar pores. To stress the oxidation process, we increased the temperature to 75°C and the residence

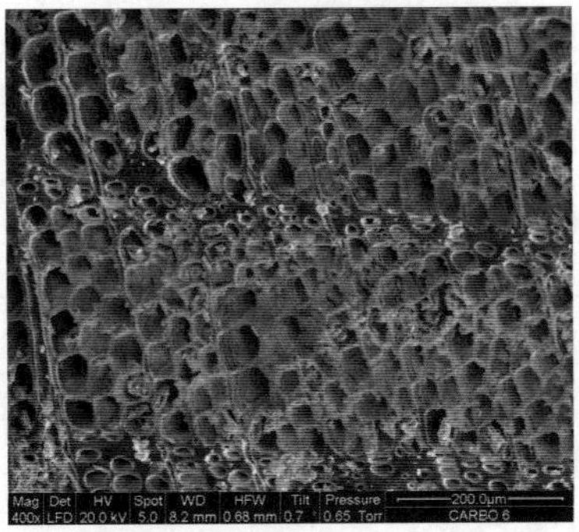

a) Old biochar comes from larch coniferous tree.

b) Modern biochar comes from larch coniferous tree

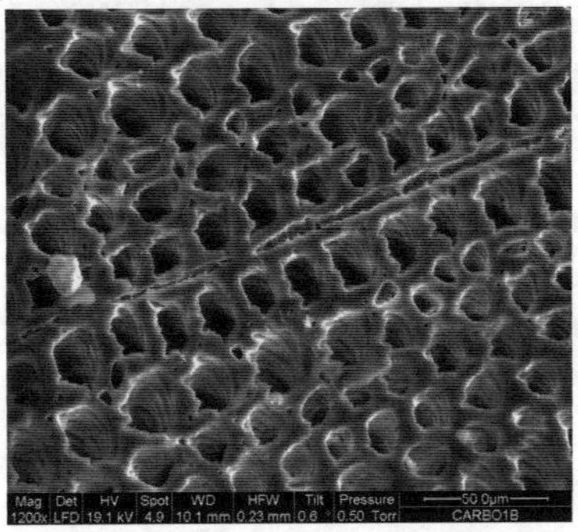

c) Old biochar comes from larch coniferous tree, on a finer scale.

d) Modern biochar comes from larch coniferous tree, on a finer scale.

Figure 2. Comparison between old (left) and modern (right) biochar by SEM technique. The pores in old biochar appear thicker than in modern biochar.

time in the thermal bath to 120 h. In figure 3 c) and d), the results relative to the SEM images are reported. Also in this case, we did not observe any significant difference in the modern biochar treated by H_2O_2.

Further investigations are forthcoming to understand the density variation between old and modern biochar and the recalcitrance of pyrogenic carbon for a long period in soil.

4. Conclusions

As a result of this analysis we can conclude that biochar stored in soil over long time scales changes its morphology. The study of the density highlights that old biochar density is almost the double of the modern biochar one. This result is confirmed by SEM analysis, where we have observed a thickening of pores inside the old biochar. Our first hypothesis assumed that the probable cause was an oxidation effect, but this is not confirmed by our experiments. Therefore, the inflation of pores is under investigation by X-ray diffraction. For instance, another hypothesis could be the presence of inorganic salts absorbed from the soil.

5. Acknowledgments

Authors acknowledge Regione Toscana, Open Fields s.r.l. and AGT s.r.l. for financial support and assistance. We thank also Dr. L. Genesio and F. P. Vaccari for scientific discussions and

a) Modern biochar without H_2O_2 treatment.

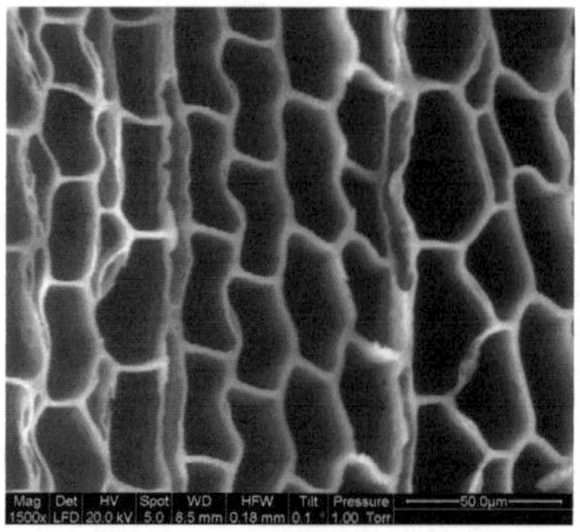

b) Modern biochar with H_2O_2 treatment for 72 h.

c) Modern biochar without H_2O_2 treatment, on a finer scale.

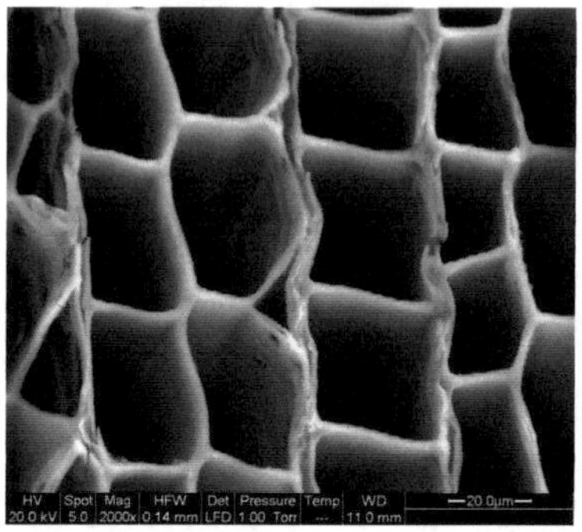

d) Modern biochar with H_2O_2 treatment for 120 h, on a finer scale.

Figure 3. a) and b) images show the modern biochar, without and with H_2O_2 treatment (35% - 130 Vol.) at 30°C for 72 h, respectively; c) and d) images show the modern biochar without and with thermal bath of H_2O_2 at about 75°C for 120 h, respectively.

encouragement during this work. This study was supported by the EuroChar project (FP7-ENV2010 ID-265179).

References

[1] Fourth Assessment Report of the Intergovernmental Panel on Climate Change, 2007. Summary for Policymakers. Available from: http://www.ipcc.ch/ipccreports/ar4-wg2.htm
[2] Woolf D, Amonette J E, Street-Perrott F A, Lehmann J and Joseph S 2010 *Nature communications* **1**, 56
[3] Baronti S, Alberti G, Delle Vedove G, Di Gennaro F, Fellet G, Genesio L, Miglietta F, Peressotti A and Vaccari F P 2009, *Italian Journal Agronomy*, Vol 5
[4] Vaccari F P, Baronti S, Lugato E, Genesio L, Castaldi S, Fornasier F and Miglietta F 2011, *European Journal of Agronomy* **34**, 231
[5] Trebbi G 1993, *Bioresource Technology* **46**, 23
[6] Yaman S 2004, *Energy Conversion and Management* **45**, 651
[7] Kaal J, Martinez Cortizas A and Nieropc K G J 2009, *Journal of Analytical and Applied Pyrolysis* **85**, 408
[8] Wood Anatomy of central European species - Microscopic Wood Anatomy of Central European species, http://www.woodanatomy.ch/
[9] Kinoshita K, Saito T, Ito A, Kawakami T, Kitagawa Y, Yamanaka S, Yamaguchi K and Okumura M 2011, *Polyhedron* **30**, 3249

4th Young Researcher Meeting, Trieste 2013 IOP Publishing
Journal of Physics: Conference Series **470** (2013) 012004 doi:10.1088/1742-6596/470/1/012004

On first attempts to reconcile quantum principles with gravity

A Rocci
Dipartimento di Fisica e Astronomia G. Galilei, Università degli Studi di Padova, I-35131 Padova (Italy)
E-mail: `rocci@pd.infn.it`

Abstract. In his first paper of 1916 on gravitational waves Einstein began to speculate on *interactions* between the principles of the old quantum theory and his theory of gravitation. With this contribution Einstein has stimulated a lot of similar speculations, during the dawn and the development of Quantum Mechanics. These speculations have culminated with the first try to quantize the gravitational field, that was provided by Rosenfeld in 1930. In this paper we briefly explain why this period (1916-1930) should be inserted into the history of Quantum Gravity and then we focus on Klein's approach to the problem of reconciling Wave Mechanics with gravity, during the two-years period 1926-1927. His attempt should be looked as the prehistory of Quantum Field Theory in a curved background.

1. Introduction

The term Quantum Gravity (QG) is often associated to the idea that the gravitational field must be quantized. Up to the present, we do not know how to construct this theory in a consistent way. From the birth of General Relativity (GR) in 1915 until today, many approaches tried to face, in a broad sense, the problem of harmonizing the quantum principles and GR.

In 1916, Einstein [1] was the first to argue that quantum effects must modify his general theory. In fact he had in mind the Bohr's principle of stationary orbits that had already modified the classical idea of the atomic collapse in the case of the energy loss due to electromagnetic wave emission and that seemed to suggest that a similar solution is needed to avoid the energy loss caused by gravitational wave emission. Presumably he did not know, at that time, that this kind of atomic collapse is characterized by a time of the order of 10^{37}s, that is an enormous lack of time compared with the life of our Universe[1] [3].

The first try to quantize the gravitational field appeared in 1930. In this year Rosenfeld published two papers. In his first paper[2] [5] Rosenfeld considers a Lagrangian that includes the electromagnetic field, the Dirac field and the gravitational field. Concerning the latter, he writes the Einstein-Hilbert action using the tetrad formalism[3]. In his approach, quantization takes place when classical variables, the tetrads, and their conjugate momenta are substituted with hermitian operators[4]. With this paper the *canonical quantization approach* enters into the

[1] Recent results [2] give approximately $(4.354 \pm 0.012) \times 10^{17}$s.
[2] See [4] for a commented English translation.
[3] Rosenfeld refers to this formalism as "the one-body theory proposed by Weyl" [6].
[4] In those years physicists were used to talk about *c-numbers* (classical) and *q-numbers* (quantum).

Journal of Physics: Conference Series **470** (2013) 012004 doi:10.1088/1742-6596/470/1/012004

history of QG. In a second paper [7], the linearization procedure for the metric $g_{\mu\nu}$, introduced by Einstein[5][1], is used: $g_{\mu\nu} = \eta_{\mu\nu} + \sqrt{\chi}h_{\mu\nu}$, where χ is proportional to the Newton constant and $h_{\mu\nu}$ is the perturbation of the Minkowski metric $\eta_{\mu\nu}$. In this article Rosenfeld writes the linearized Lagrangian and chooses, as field variables, the perturbation of the metric, that he rewrites in terms of annihilation and creation operators. This splitting procedure will be adopted also later on by Gupta [8] and Feynman [9], the fathers of the *covariant quantization approach.*

Now we want to raise the following question: should the endeavours *preceding* Rosenfeld's work be considered part of the history of QG? To answer the question we point out the following facts. First of all, the idea that *every field can be quantized* appeared for the first time in 1929, thanks to Heisenberg and Pauli [10]. Despite this, Einstein had already speculated on possible interactions between the principles of the old quantum theory[6] and his theory of gravitation. Likewise, Klein[7] and other researchers, during the early years of Quantum Mechanics (QM), tried to find out a unifying framework that could harmonize Wave Mechanics[8] with gravity. It was a failed[9] attempt to modify GR for reconciling it with quantum principles and some authors have already underlined this fact [13]-[15]. For this reason we decided to study more in detail Klein's program from the perspective of the history of QG. To answer the question we posed, another important fact to consider is the role played by the semi-classical methods born after Rosenfeld papers, like e.g. Quantum Field Theory (QFT) in curved space-times. Even though they are not directly connected with QG problems, they revealed new phenomena and raised new problems that will find a solution in a consistent theory of QG. We have in mind the black hole radiation [16] and the *information loss paradox* [17], discovered by Hawking in the seventies. Concerning the latter, it will be avoided by a unitary description of the black hole radiation process in the framework of a consistent theory of QG. From a modern point of view, QFT in curved backgrounds is a *semi-classical* approach, because *quantum* matter fields live in a *classical* curved background: we can consider this approach as a way to reconcile quantum principles with gravity at an energy scale where the quantum effects of the gravitational field are negligible and for this reason we believe that the development of these methods should belong to the history of QG. And last but not least, let us consider String Theory, one of the most promising approach to QG. It combines together various features coming from different areas, like e.g. Klein's idea of extra dimensions or QFT in two dimensions, without quantizing the gravitational field directly. In fact String Theory offers a picture where QG *emerges* from the quantum theory of strings [18].

For these reasons we think that *every* endeavour to reconcile quantum principles with gravity should be inserted in the history of QG, even though in the approach considered the gravitational field is treated classically. We imagine the history of QG as a chain of islands; on every island there is a farm, with physicists at work, and every island corresponds to a different approach to the problem of reconciling quantum postulates with the theory of gravitation.

Our paper aims to consider a particular farm, that tried to unify gravity, Maxwell's theory and Wave Mechanics, with the main focus on Klein's work in the years 1926 and 1927. As far as we know, we note for the first time that Klein's contribution should be inserted into the *prehistory* of QFT in curved backgrounds.

[5] We use the signature $(-,+,+,+)$ and μ, ν run from 0 to 3.

[6] This term refers, as usual, to all models developed to explain new phenomena that arose between 1900 and 1925, using quantized classical quantities (Planck's black body theory, Einstein-Bohr-Sommerfeld quantization rules etc.).

[7] We refer to Oskar Klein, not to be confused with Felix Klein.

[8] In the old quantum theory there were two main approaches [11]: the *particle* approach and the *wave* approach. They led to the birth of Matrix Mechanics in 1925, developed by Heisenberg, Born and Jordan, and to the birth of Wave Mechanics at the beginning of 1926, developed by Schrödinger.

[9] Klein himself admitted that his program failed because "as Dirac may well say, my main trouble came from trying to solve too many problems at a time!"[12].

Journal of Physics: Conference Series **470** (2013) 012004 doi:10.1088/1742-6596/470/1/012004

2. Klein's contribution

Maxwell's legacy has shown that unifying different areas of physics could lead to a better understanding of our world. Following this idea, some researchers started to search for a *unified theory*, that could describe both GR and Maxwell's theory in a unified framework. Among these, some physicists assumed that we live in a *five-dimensional* world: in addition to the usual four space-time dimensions, there is a fifth space-like dimension[10], x^5, that is compactified on a circle[11] of radius l. In this approach, the five-dimensional metric $\gamma_{\bar{\mu}\bar{\nu}}$ $\bar{\mu},\bar{\nu} = 0,1,2,3,5$, once decomposed in its low-dimensional components, can describe gravity and Maxwell's theory. In fact $\gamma_{\mu\nu}$ and $\gamma_{5\mu}$ behave like a four-dimensional symmetric tensor and a four-dimensional vector respectively and then they could play the role of the metric and of the electromagnetic potential. Kaluza and Klein are known as the fathers of this approach[12].

Kaluza made the assumption of cilindricity in 1921 [19]. In the last sentence of his paper he declares to be afraid that quantum theory would be a menace for his theory[13]. Klein found out the same approach later and independently in 1926. On the contrary he was convinced that quantum principles should play a fundamental role in his version of a five-dimensional unified theory.

Klein's program can be summarized in the following three steps: writing a five-dimensional wave equation[14] for a massive particle (electron/proton); embedding it into a curved space-time; getting a conservation principle. In his first paper [20], he introduces two five-dimensional metrics, related to each other and depending on the four-dimensional x^μ only. The first line element is $d\sigma^2 = \gamma_{\bar{\mu}\bar{\nu}}(x^\mu)dx^{\bar{\mu}}dx^{\bar{\nu}}$ that is decomposed in the following way:

$$\begin{aligned} d\sigma^2 &= \gamma_{55}(dx^5)^2 + 2\gamma_{5\mu}dx^5dx^\mu + \gamma_{\mu\nu}dx^\mu dx^\nu = \gamma_{55}\left(dx^5 + \frac{\gamma_{5\mu}}{\gamma_{55}}dx^\mu\right)^2 + \left(\gamma_{\mu\nu} - \frac{\gamma_{5\mu}\gamma_{5\nu}}{\gamma_{55}}\right)dx^\mu dx^\nu \\ &= \alpha\,(d\vartheta)^2 + g_{\mu\nu}dx^\mu dx^\nu = \alpha d\vartheta^2 + ds^2. \end{aligned} \tag{1}$$

Klein decided to set $\alpha = \frac{16\pi G}{e^2c^2}$, where G, c and e are the Newton constant, the speed of light and the electric charge respectively. Using these Ansätze, Klein proved that the variational principle applied to an Einstein-Hilbert action in five dimensions leads to a set of equations that can be identified with the four-dimensional Einstein's equations and the Maxwell's equations[15]:

$$\delta\int d^5x\sqrt{-\gamma}R^{(5)} = 0 \quad\rightarrow\quad R^{(4)\mu\nu} - \frac{1}{2}g^{\mu\nu}R^{(4)} = \frac{8\pi G}{c^4}T^{\mu\nu} \quad , \quad \partial_\mu\left(\sqrt{-g}F^{\mu\nu}\right) = 0$$

This happens because Klein identifies the $g_{\mu\nu}$ of equation 1 with the four-dimensional metric and because he defines the electromagnetic four-potential A_μ, whose field strength is $F^{\mu\nu}$, in the following way:

$$A_\mu = \frac{c}{e}\left(\frac{\gamma_{5\mu}}{\gamma_{55}}\right) \qquad F_{\mu\nu} = \partial_\mu A_\nu - \partial_\nu A_\mu.$$

The second line element that he introduces is $d\hat{\sigma}^2 = a_{\bar{\mu}\bar{\nu}}dx^{\bar{\mu}}dx^{\bar{\nu}}$ defined by:

$$d\hat{\sigma}^2 = \hat{k}d\vartheta^2 + ds^2 = d\sigma^2 + \left(\hat{k} - \alpha\right)d\vartheta^2$$

[10] Because of this assumption, from now on, we will call this approach *the five-dimensional farm.*

[11] This Ansatz is often known as the hypothesis of *cilindricity.*

[12] The γ_{55} component behaves like a scalar field and it is today known as the *dilaton.* Its role was not fully understood at that time: it was set as a constant by these authors.

[13] "After all, what threatens all the Ansatz, which demand universal validity, is the sphinx of modern physics - quantum theory"

[14] As we will see, the wave equation he used is the *Klein-Gordon* (K-G) *equation.*

[15] In this sense, Klein's theory unifies gravitational and electromagnetic forces. $T^{\mu\nu}$ is the usual electromagnetic energy-momentum tensor.

As stated above, in 1925 Klein was convinced that a starting point to reconcile QM with GR was to introduce a relativistic version of the Schrödinger equation in a curved background. As a first try, Klein followed the analogy with light[16], then he decided to set $\hat{k} = \frac{1}{m^2c^2}$, where m represents the electron mass, to hide it into the previous metric and finally he wrote a five-dimensional *massless* wave equation[17]:

$$a^{\bar{\mu}\bar{\nu}} \left(\frac{\partial}{\partial x^{\bar{\mu}}} + \Gamma^{\bar{\sigma}}_{\bar{\mu}\bar{\nu}} \right) \partial_{\bar{\sigma}} \Psi = a^{\bar{\mu}\bar{\nu}} D_{\bar{\mu}} \partial_{\bar{\nu}} \Psi = 0. \tag{2}$$

The *covariant derivative* $D_{\bar{\mu}}$ tells us that the dynamic is on a curved manifold[18] and the equation reproduces, in the optical geometric limit, an Hamilton-Jacobi equation for a light-like wave. From a modern point of view, this is the wave equation for a massless scalar field.

Using again the metric $a_{\bar{\mu}\bar{\nu}}$ Klein introduces a Lagrangian, that represents the dynamic of a *massless* particle in five-dimensions:

$$L = \frac{1}{2} a_{\bar{\mu}\bar{\nu}} \frac{dx^{\bar{\mu}}}{d\lambda} \frac{dx^{\bar{\nu}}}{d\lambda} = \frac{1}{2} \left[\hat{k} \left(\frac{d\vartheta}{d\lambda} \right)^2 + \left(\frac{ds}{d\lambda} \right)^2 \right]$$

where λ is the affine parameter that parametrizes the geodesics. The conjugated momenta $p^{\bar{\mu}}$, defined as usual by $p^{\bar{\mu}} = \frac{\partial L}{\partial(dx^{\bar{\mu}}/d\lambda)}$, produce a conservation principle, $\frac{dp^5}{d\lambda} = 0$ because the metric does not depend on x^5, and the four-dimensional geodesic equation for a *charged electron* moving in a gravitational and electromagnetic field.

The unusual connection between the *massless* wave equation and the *massive* electron was noted by de Broglie at the beginning of 1927 [21] and in his last paper [22] Klein rewrites equation 2 in an equivalent form, using the inverse of the *first* metric we introduced, $\gamma^{\bar{\mu}\bar{\nu}}$:

$$\gamma^{\bar{\mu}\bar{\nu}} D_{\bar{\mu}} \partial_{\bar{\nu}} \Psi = \mathcal{I}^2 \Psi \qquad \mathcal{I}^2 = \frac{1}{\hbar^2} \left(m^2c^2 - \frac{e^2c^2}{16\pi G} \right). \tag{3}$$

From our point of view, this is the K-G equation for a *massive* scalar field living in a curved background.

In particular, in this equation it appears the Planck constant[19] $\hbar$. To understand this fact, we start noting that the momentum p^5 does not have the usual dimension, because α, and x^5 as a consequence, is not dimensionless[20]. This means that the fifth coordinate having the right length dimension is $\tilde{x}^5 = \sqrt{\alpha} x^5$. Relaxing completely the analogy with light, Klein introduces another Lagrangian [23] using the *first* metric:

$$\tilde{L} = \frac{1}{2} m \gamma_{\bar{\mu}\bar{\nu}} \frac{dx^{\bar{\mu}}}{d\lambda} \frac{dx^{\bar{\nu}}}{d\lambda}.$$

This is the Lagrangian for a *massive* particle in five dimensions and the fifth component of the momentum associated to $\tilde{x}^5$ is conserved again: $\frac{d\tilde{p}^5}{d\lambda} = 0$. This means that $\tilde{p}^5$ is a constant to be determined. As before, if we ask to the four-momentum p^μ to reproduce the geodesic equation, the following identity must be satisfied: $\sqrt{\alpha}\tilde{p}^5 = 1$. Using the Ansatz of cilindricity, Klein writes $\tilde{p}^5 = \frac{nh}{l}$, as usual in QM, and then he sets $n = 1$. The wave function Ψ must be periodic in the

[16] The analogy's tool has already guided Klein in introducing the five-dimensional space-time [12].
[17] For scalar functions a covariant derivative is equivalent to an ordinary derivative.
[18] The Cristoffel symbols $\Gamma^{\bar{\sigma}}_{\bar{\mu}\bar{\nu}}$ are related to the metric $\gamma^{\bar{\mu}\bar{\nu}}$.
[19] De Broglie was proud of the fact that this equation contains *all* known constants of Nature.
[20] See equation 1

fifth coordinate and equating both expression for $\bar{p}^5$ we get the period: $l = \sqrt{\alpha}h$. Inserting this conditions into equation 2 makes the Planck constant to appear. From Klein's point of view, these equalities give a connection between the quantization rules and the quantization of the electric charge. But the only consequence of that is an estimation for the radius of the fifth dimension: the Planck length. Klein did not realize that setting $n = 1$ implies to have a particle with mass equal to the Planck mass, as it happens today in String Theory [18].

In his last paper of the two-year period [22], Klein follows his program again, but looking for a more general approach. In fact the starting point is the following invariant, that is proportional to the action for a massive scalar field[21]:

$$\mathcal{S} = \int \sqrt{-\gamma}\,\mathcal{L}\,d^5x = \int \sqrt{-\gamma}\,\frac{\hbar^2}{m}\left[\frac{1}{2}\partial_{\bar{\mu}}\Psi\partial^{\bar{\mu}}\Psi + \frac{1}{2}\mathcal{I}^2\Psi^2\right] d^5x. \tag{4}$$

Now, varying the action with respect to the dynamical variables Ψ and $\gamma_{\bar{\mu}\bar{\nu}}$, he obtains a new five-dimensional *conservation principle*. In fact we can write:

$$\delta\mathcal{S} = \delta_\gamma\mathcal{S} + \delta_\Psi\mathcal{S} = \int \sqrt{-\gamma}\left[\Theta_{\bar{\mu}\bar{\nu}}\delta\gamma^{\bar{\mu}\bar{\nu}} + \frac{\hbar^2}{m}\left(D_{\bar{\mu}}\partial^{\bar{\mu}}\Psi - \mathcal{I}^2\Psi\right)\delta\Psi\right] d^5x$$

where $\Theta_{\bar{\mu}\bar{\nu}}$ is one half of the energy-momentum tensor of the scalar field as we define it today [24]. Introducing an infinitesimal variation of the coordinates $x^{\bar{\mu}} \to x^{\bar{\mu}} + \xi^{\bar{\mu}}$ the fields vary as usual, $\delta\Psi = \partial_{\bar{\mu}}\Psi\xi^{\bar{\mu}}$ and $\delta\gamma_{\bar{\mu}\bar{\nu}} = D_{(\bar{\mu}}\xi_{\bar{\nu})}$, and integrating by parts we get:

$$\delta\mathcal{S} = \int \sqrt{-\gamma}\left[D^{\bar{\mu}}\Theta_{\bar{\mu}\bar{\nu}} + \frac{\hbar^2}{m}\left(D_{\bar{\mu}}\partial^{\bar{\mu}}\Psi - \mathcal{I}^2\Psi\right)\partial_{\bar{\nu}}\Psi\right]\xi^{\bar{\nu}}d^5x$$

At the end, imposing the variational principle and the wave equation for Ψ in the last equation, Klein obtains the covariant version of the energy-momentum conservation law[22]: $D^{\bar{\mu}}\Theta_{\bar{\mu}\bar{\nu}} = 0$.
We would like to conclude this section with some remarks.

As far as we know, the action for a scalar field, the related energy-momentum tensor and its conservation law, connected with the free equations, appear for the *first* time in Klein's papers. Klein never used the word *field* referring to the Ψ, he always explicitly considers it as the wave function of the electron. In spite of this, his approach is similar to the point of view that will be adopted by developers of QFT in curved backgrounds.

In 1927 Klein and Jordan [25] published the paper where they connect the Bose-Einstein statistics to the K-G equation. Despite this, at that time Klein considered Ψ like the wave function associated to the electron. Pais [26] pointed out that, in the Jordan-Klein paper, the authors treated the wave function as a field, in fact they write Ψ in terms of annihilation and creation operator. Nonetheless, in his subsequent paper on five-dimensional approach, the last we commented, Klein never use this decomposition.

We can appreciate how Klein changed his point of view, during the two-year period. In the Concluding Remarks of the first paper of the period [20], he presents as fundamental his introduction of the fifth dimension. In particular he states that it is possible to understand this radical modification "through quantum theory". At the end of the introduction of his last paper [22], Klein realizes that his approach is incomplete, but he is still convinced that his procedure is a natural starting point to construct a general theory of quantum fields[23].

[21] The action did not have the meaning that we give it today in field theory and an overall minus sign is missing in front of the action.
[22] He also applies this procedure to the free electromagnetic field case, where it works in a similar way.
[23] "Hierdurch und nach dem erwähnten Gesichtpunkt scheint sich diese fünfdimensionale Form der Relativitätstheorie als der natürlische Ausgangspunkt für eine allgemeine Quantenfeldtheorie darzubieten".

3. Conclusions

In the years 1926-1927 the five-dimensional farm was an active research area. All physicists belonging to the farm tried to modify GR to reconcile it with Wave Mechanics. From our modern point of view they lay in an intermediate stage between the first and the second quantization approach. In Klein's papers we find for the first time the action, the equations of motion and the energy-momentum tensor for a massive scalar field living in a curved background. His starting point will be shared by developers of QFT in curved backgrounds. Even though they are not directly connected with QG, these attempts should belong to the history of QG. In particular we suggest to insert Klein's contribution in the *prehistory* of QFT in curved backgrounds.

Acknowledgments

We are grateful to Prof. G. Peruzzi and Prof. K. Lechner for the patience they always demonstrate, for fruitful discussions and for their valuable suggestions. We are definitely indebted to Sara for her loving and fundamental support.

References

[1] Einstein A 1916 Närungsweise Integration der Feldgleichungen der Gravitation *Preussische Akademien der Wissenschaften. Sitzungsberichte* (Berlin) 688
[2] Planck Collaboration (Ade P A R *et al.*) 2013 Planck 2013 results. I. Overview of products and scientific results *Preprint* astro-ph.CO/1303.5062
[3] Gorelick G E *The first steps of quantum gravity and the Planck values* Studies in the History of General Relativity (Einstein Studies vol 4) ed J Eisenstaedt and A J Kox (Boston/Basel/Berlin: Birkäuser)
[4] Salisbury D 2009 Translation and commentary of Léon Rosenfeld's 'Zur Quantelung der Wellenfelder', Annalen der Physik 397, 113 (1930) *Preprint* http://www.mpiwg-berlin.mpg.de/en/resources/preprints.html
[5] Rosenfeld L 1930 Zur Quantelung der Wellenfelder *Annalen der Physik* **5** 1113
[6] Weyl H 1929 Elektron und Gravitation I *Zeitschrift für Physik* **56** 330
[7] Rosenfeld L 1930 Über die Gravitationswirkungen des Lichtes *Zeitschrift für Physik* **65** 589
[8] Gupta S N 1952 Quantization of Einstein's gravitational field: linear approximation *The Proceedings of the Physical Society* **A65** 161
[9] Feynman R P 1963 Quantum theory of gravitation *Acta Physica Polonica* **24** 697
[10] Heisenberg W e Pauli W 1929 Zur Quantendynamik der Wellenfelder *Zeitschrift für Physik* **56** 1
[11] Pais A 1988 *Inward Bound: Of Matter and Forces in the Physical World* (Oxford: Oxford University Press)
[12] Klein O 1991 *From my Life of Physics* (The Oskar Klein Memorial Lectures vol 1) ed G Ekspong (Stokholm: Stockholm University)
[13] Goenner H F M 2004 On the history of unified field theories *Living Rev. Relativity* **7** 2 http://www.livingreviews.org/lrr-2004-2
[14] O'Raifeartaigh L 1997 *The dawning of Gauge Theory* (Princeton: Princeton Series in Physics)
[15] O'Raifeartaigh L and Straumann N 2000 Gauge theory: historical origins and some modern developments *Rev.Mod.Phys.* **72** 1
[16] Hawking S W 1975 Particle creation by black holes *Comm. Math. Phys.* **43** 199
[17] Hawking S W 1976 Breakdown of predictability in gravitational collapse *Phys. Rev.* D **14** 2460
[18] Polchinski J 2001 *String Theory. Vol 1: an Introduction to the Bosonic String* (Cambridge: Cambridge University Press)
[19] Kaluza T 1921 Zum Unitätsproblem der Physik *Sitzungsberichte Preussische Akademie der Wissenschaften* **96** 69
[20] Klein O 1926 Quantentheorie und fünfdimensionale Relativitätstheorie *Zeitschrift für Physik* **37** 895
[21] De Broglie L 1927 L'univers a cinq dimensions et la mécanique ondulatoire *Le Journal de Physique et le Radium* **s. VI t. VIII** N^o**2** 65
[22] Klein O 1927 Zur fünfdimensionalen Darstellung der Relativitätstheorie *Zeitschrift für Physik* **46** 188
[23] Klein O 1926 The atomicity of electricity as a quantum theory law *Nature* **118** 516
[24] De Felice F and Clarke C J S 1992 *Relativity on curved manifolds* (Cambridge: Cambridge University Press)
[25] Jordan P and Klein O 1927 Zum Mehrkörperproblem der Quantentheorie *Zeitschrift für Physik* **45** 751
[26] Pais A 2000 *The Genius of science: a portrait gallery of twentieth-century physicists* (New York: Oxford University Press)

Journal of Physics: Conference Series **470** (2013) 012005 doi:10.1088/1742-6596/470/1/012005

Detection methods to rule out completely co-positive and bi-entangling operations

C Macchiavello and M Rossi[†]

Dipartimento di Fisica and INFN-Sezione di Pavia, via Bassi 6, 27100 Pavia, Italy

E-mail: [†] matteo.rossi@unipv.it

Abstract. In this work we extend the quantum channel detection method developed in Refs. [1, 2] in order to detect other interesting convex sets of quantum channels. First we work out a procedure to detect non completely co-positive maps. Then we focus on the set of so-called bi-entangling operations and show how a map outside this set can be revealed. In both cases we provide explicit examples showing the theoretical technique and the corresponding experimental procedure.

1. Introduction

In quantum information it is of great importance to characterise quantum communication channels or quantum devices without necessarily performing quantum process tomography. Actually, quantum process tomography requires a large number of experimental resources, while one is usually interested in few properties of the quantum channel under consideration, as e.g. whether the channel has some entangling power. In many realistic implementations some a priori information on the form of the channel is available, hence, the quantum channel detection (QCD) method developed in [1, 2] can be applied. Besides being less informative than the full quantum process tomography, the QCD method allows us to test the property of interest with a much smaller experimental effort.

In this work we will discuss in detail how to detect two sets of quantum channels, namely quantum channels that are not completely co-positive (CCOP) and the set of operations that are not bi-entangling (BE). Both sets are of great interest as they are connected via the Choi-Jamiolkowski isomorphism to PPT states[1] [3] and to the problem of classical simulatability of quantum computation [4], respectively.

This work is organized as follows. In section 2 we will review the main idea of QCD following Refs. [1, 2]. In section 3 we will discuss a method to detect non CCOP maps. In section 4 we will study how to reveal quantum channels that are not BE operations, and we finally summarize the main results in section 5.

2. The general QCD method

The QCD method proposed in Refs. [1, 2] relies on the concept of witness operators [5] and the Choi-Jamiolkowski isomorphism [6]. We briefly remind both of them in the following.

[1] A state of a bipartite system is PPT if the partial transpose of its density matrix is positive semi-definite, otherwise it is NPT.

Journal of Physics: Conference Series **470** (2013) 012005 doi:10.1088/1742-6596/470/1/012005

A state ρ is entangled if and only if there exists a hermitian operator W such that $\mathrm{Tr}[W\rho] < 0$ and $\mathrm{Tr}[W\rho_{sep}] \geq 0$ for all separable states; such an operator is called an entanglement witness. The Choi-Jamiolkowski isomorphism provides a one-to-one correspondence between completely positive (CP) maps $\mathscr{M}$ acting on $\mathcal{D}(\mathcal{H})$ (the set of density operators on $\mathcal{H}$, with finite dimension d) and positive operators $C_{\mathscr{M}}$ on $\mathcal{H}_A \otimes \mathcal{H}_B$ (named Choi states), where A and B here denote the two subsystems on which the Choi state is defined. The isomorphism can be stated as

$$\mathscr{M} \Longleftrightarrow C_{\mathscr{M}} = (\mathscr{M} \otimes \mathscr{I})[|\alpha\rangle\langle\alpha|], \tag{1}$$

where $\mathscr{I}$ is the identity map, and $|\alpha\rangle$ is the maximally entangled state with respect to the bipartite space $\mathcal{H} \otimes \mathcal{H}$, i.e. $|\alpha\rangle = \frac{1}{\sqrt{d}} \sum_{k=1}^{d} |k\rangle\,|k\rangle$. The above isomorphism can be exploited to link convex sets of quantum channels to particular convex sets of quantum states.

As a simple example consider the convex set of entanglement breaking (EB) channels. A channel $\mathscr{E}$ is EB if and only if its Choi state $C_{\mathscr{E}}$ is separable [7]. Therefore, the detection of entanglement of $C_{\mathscr{E}}$ by means of a suitable witness operator W_{EB} implies that the implemented quantum channel $\mathscr{E}$ is not EB [1].

Although the general QCD method applies to several convex sets of quantum channels, as e.g. EB and separable maps, in the following it will be explicitly studied for the convex sets of CCOP channels and BE operations.

3. Completely co-positive channels

In this section we will consider the set of CCOP channels. A CP map $\mathscr{C}$ acting on a qudit (d-dimensional system) is CCOP if and only if the composite map $\mathscr{C}_{\mathscr{T}} = \mathscr{T} \circ \mathscr{C}$, where $\mathscr{T}$ is the transposition map, is CP. Since a quantum map is CP if and only if the corresponding Choi operator is positive, we can restate the above definition as follows: a CP map $\mathscr{C}$ is CCOP if and only if the Choi operator $C_{\mathscr{C}_{\mathscr{T}}}$ related to the composite map $\mathscr{C}_{\mathscr{T}}$ is positive.

By the above correspondence we will develop a method to detect whether a map is non CCOP by adapting techniques developed for the detection of non positive partial transposed (NPT) entangled states [3]. Consider then a map $\mathscr{M}$ that does not belong to the set of CCOP channels. From the above definition it follows that the bipartite Choi state $C_{\mathscr{M}_{\mathscr{T}}} = (\mathscr{T}_A \otimes \mathscr{I})[C_{\mathscr{M}}]$ has at least one negative eigenvalue. Let λ_- be the most negative eigenvalue corresponding to the eigenvector $|\lambda_-\rangle$. The following operator, i.e.

$$W_{\mathrm{CCOP}} = |\lambda_-\rangle\langle\lambda_-|^{\mathscr{T}_A}, \tag{2}$$

is thus suitable to detect the NPT state $C_{\mathscr{M}_{\mathscr{T}}}$ corresponding to the non CCOP map $\mathscr{M}_{\mathscr{T}}$. Notice that the transposition map on the Choi state acts only on the first qudit, i.e. $\mathscr{T}_A$.

As an illustrative example we consider the case of the dephasing noise $\mathscr{D}$ acting on a single qubit, defined by the following trace preserving CP map

$$\mathscr{D}[\rho] = p\rho + (1-p)\sigma_z \rho \sigma_z \,, \tag{3}$$

where σ_z is a Pauli operator[2]. It is easy to see that the Choi state $C_{\mathscr{D}}$ corresponding to $\mathscr{D}$ takes the form

$$C_{\mathscr{D}} = p|\alpha\rangle\langle\alpha| + (1-p)|\phi^-\rangle\langle\phi^-|, \tag{4}$$

[2] Hence, dephasing noise consists in either leaving the input state unchanged (with probability p) or applying a phase flip σ_z (with $1-p$). Generally speaking, it represents a loss of quantum coherence in the off-diagonal terms of the regarded system.

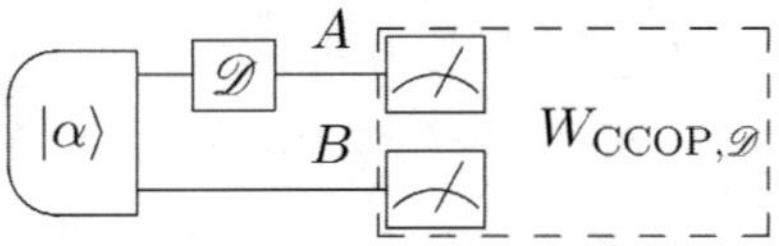

Figure 1. Experimental scheme showing the detection of the dephasing channel $\mathscr{D}$ as a non CCOP channel. Notice that the expectation value of the witness $W_{\mathrm{CCOP},\mathscr{D}}$, namely $\mathrm{Tr}[W_{\mathrm{CCOP},\mathscr{D}} C_{\mathscr{D}}]$, can be measured locally.

with $|\phi^-\rangle = \frac{1}{\sqrt{2}}(|00\rangle - |11\rangle)$. The above state can be shown to be NPT whenever $p \neq 1/2$. It is then possible to derive the following detection operator [8, 9] from Eq. 2:

$$W_{\mathrm{CCOP},\mathscr{D}} = \begin{cases} \frac{1}{4}(\mathbb{1} \otimes \mathbb{1} + \sigma_x \otimes \sigma_x + \sigma_y \otimes \sigma_y - \sigma_z \otimes \sigma_z) & \text{for } p < \frac{1}{2}, \\ \frac{1}{4}(\mathbb{1} \otimes \mathbb{1} - \sigma_x \otimes \sigma_x - \sigma_y \otimes \sigma_y - \sigma_z \otimes \sigma_z) & \text{for } p > \frac{1}{2}. \end{cases} \quad (5)$$

This method can be experimentally implemented by preparing a two-qubit state in the maximally entangled state $|\alpha\rangle$, then operating with the quantum channel $\mathscr{D}$ to be detected on one of the two qubits and measuring the operator $W_{\mathrm{CCOP},\mathscr{D}}$ acting on both qubits at the end (see Fig. 1). If the resulting average value $\mathrm{Tr}[W_{\mathrm{CCOP},\mathscr{D}} C_{\mathscr{D}}]$ is negative, we can then conclude that the Choi state $C_{\mathscr{D}_{\mathscr{T}}} = \mathscr{T} \circ \mathscr{D}$ is NPT and that the channel under consideration is not CCOP.

Finally, we would like to stress that, since every NPT state is entangled, the detection of a non CCOP channel $\mathscr{M}$ implies that $\mathscr{M}$ is not EB as well, but the opposite does not hold in general. Actually the set of EB channels is in general a subset of the CCOP channels. The two sets coincide only when the channels act on two-dimensional systems. From the perspective of QCD, this implies that for higher dimensional systems a quantum channel which is detected as non EB may nevertheless belong to the set of CCOP maps.

4. Bi-entangling operations

In this section we will focus on BE operations, a class of quantum channels that can generate at most bipartite entanglement. They were introduced in Ref. [4] in the context of quantum computation and were shown to be efficiently simulatable classically. BE operations are quantum channels acting on bipartite systems AB (of finite dimension d) in such a way that they can be expressed as convex combinations of (a) separable operations, (b) operations that swap the two qudits and then act as a separable operation, and (c) EB channels, that break any entanglement between the two qudits on which the channel acts and extra ancillae [4]. Via the Choi-Jamiolkowski isomorphism we can then characterize the set of BE operations in terms of the corresponding Choi states.

Consider a BE operation $\mathscr{M}_{BE}$ acting on the bipartite system AB. The Choi state $C_{\mathscr{M}_{BE}}$ associated to $\mathscr{M}_{BE}$ is then a four-partite state (composed of subsystems A, B, C and D). Separable channels have separable Choi states with respect to the bipartition $AC|BD$ [10]. As a consequence, channels of type (b), with a swap gate followed by separable channels, have separable Choi states in $AD|BC$. EB channels correspond to separable Choi states in the bipartition $AB|CD$. A general Choi state $C_{\mathscr{M}_{BE}}$ for a BE channel can then be written as a convex combination of four-partite states biseparable with respect to bipartitions $AC|BD$, $AD|BC$ and $AB|CD$, namely

$$C_{\mathscr{M}_{BE}} = p \sum_i p_i C_i^{(AC|BD)} + q \sum_j q_j C_j^{(AD|BC)} + r \sum_k r_k C_k^{(AB|CD)}, \quad (6)$$

where (p, q, r), $\{p_i\}$, $\{q_j\}$ and $\{r_k\}$ are probability distributions. Notice that the first term corresponds to the set (a), the second to (b) and the third to (c). In other words, the Choi states $C_{\mathscr{M}_{BE}}$ corresponding to BE operations lie in the convex hull of states biseparable with respect to the bipartitions $AC|BD$, $AD|BC$ and $AB|CD$ for the four-partite system $ABCD$. We name this convex set of four-partite Choi states corresponding to BE operations as S_{BE}. It is now possible to develop detection procedures for BE operations by employing suitable witness operators that detect the corresponding Choi state with respect to the biseparable states belonging to S_{BE}.

We will now focus on the case of a unitary transformation U acting on two d-dimensional systems. The corresponding Choi state is pure and given by $|U\rangle = (U \otimes \mathbb{1})|\alpha\rangle$. Therefore a suitable detection operator for U as a non BE operation can be constructed as [1, 2]

$$W_{BE,U} = \alpha_{BE}^2 \, \mathbb{1} - C_U \, , \tag{7}$$

where $C_U = |U\rangle\langle U|$, and the coefficient α_{BE} is the overlap between the closest biseparable state in the set S_{BE} and the entangled state $|U\rangle$, namely

$$\alpha_{BE}^2 = \max_{\mathscr{M}_{BE}} \langle U| C_{\mathscr{M}_{BE}} |U\rangle \, . \tag{8}$$

Since the maximum of a linear function over a convex set is always achieved on the extremal points, the maximum involved in α_{BE} can be always calculated by maximising over the pure biseparable states in S_{BE}, i.e.

$$\alpha_{BE} = \max_{|\Xi\rangle \in S_{BE}} |\langle \Xi | U\rangle| . \tag{9}$$

By exploiting the Schmidt decomposition [11] of the state $|U\rangle$, the maximization above can be expressed analytically as

$$\alpha_{BE} = \max_i \max_\lambda \lambda_i(U), \tag{10}$$

where the index i labels the bipartitions $AC|BD$, $AD|BC$ and $AB|CD$, and $\lambda_i(U)$ are the Schmidt coefficients of $|U\rangle$ in the bipartition i. Therefore, in order to find the coefficient α_{BE} one has to find the maximal Schmidt coefficient of $|U\rangle$ for a fixed bipartite splitting and then maximize it among all the bipartitions involving only two versus two subsystems.

As an example of the above procedure consider the following unitary operation V acting on a two-qubit system

$$V = \begin{pmatrix} 1 & 0 & 0 & 0 \\ 0 & 0 & 1 & 0 \\ 0 & 1 & 0 & 0 \\ 0 & 0 & 0 & -1 \end{pmatrix} \tag{11}$$

The gate V is a modified swap gate such that it is no longer a BE operation. The coefficient α_{BE} for V can be computed following the steps outlined above. The Choi state $|V\rangle$ associated to the gate V is given by

$$|V\rangle = \frac{1}{\sqrt{2}}(|\alpha\rangle_{AD} |00\rangle_{BC} + |\phi^-\rangle_{AD} |11\rangle_{BC}), \tag{12}$$

and the Schmidt coefficients of V with respect to the bipartitions $AC|BD$, $AD|BC$ and $AB|CD$ can be easily computed as $\lambda_{AC|BD}(V) = (\frac{1}{2}, \frac{1}{2}, \frac{1}{2}, \frac{1}{2})$, $\lambda_{AD|BC}(V) = (\frac{1}{\sqrt{2}}, \frac{1}{\sqrt{2}}, 0, 0)$ and $\lambda_{AB|CD}(V) = (\frac{1}{2}, \frac{1}{2}, \frac{1}{2}, \frac{1}{2})$. Therefore, the coefficient α_{BE} equals $1/\sqrt{2}$ and a suitable detection operator in order to detect V as a non BE operation takes the form

$$W_{BE,V} = \frac{1}{2} \mathbb{1} - C_V . \tag{13}$$

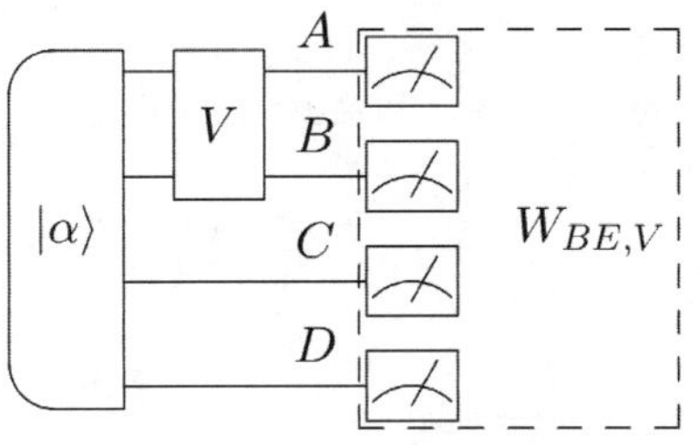

Figure 2. Experimental scheme implementing the detection of the gate V defined in Eq. 11 as a non BE operation.

From an experimental point of view, the detection procedure can be implemented as follows: prepare a four-partite qubit system in the state $|\alpha\rangle = |\alpha\rangle_{AC} |\alpha\rangle_{BD}$, apply the quantum gate V to qubits A and B, and finally perform a suitable set of local measurements in order to measure the operator (13). If the resulting average value $\mathrm{Tr}[W_{BE,V}C_V]$ is negative then the quantum channel is detected as a non BE operation. The experimental scheme is shown in Fig. 2.

We conclude this section by noticing that the method described above leads to different detection operators with respect to the detection method for non separable maps [1]. Indeed, already in the case of two qubits, the optimal witness operator that detects the gate V as a non separable channel is [1]

$$W_{Sep,V} = \frac{1}{4}\mathbb{1} - C_V. \tag{14}$$

As expected, the detection operator above is weaker than $W_{BE,V}$ in Eq. 13 in the sense that it leads to a negative expectation value for a smaller set of CP maps. This is due to the fact that BE maps are a strict subset of separable maps, and actually the set of separable Choi states in the bipartition $AC|BD$, corresponding to separable maps, is a strict subset of S_{BE}.

5. Conclusions

In summary, after a brief review of the general quantum channel detection method proposed in [1], based on the Choi-Jamiolkowski isomorphism and witness operators, we have developed a method to detect maps that do not belong to specific convex sets, i.e. the completely co-positive maps and the bi-entangling operations. Significant examples of non co-positive and non bi-entangling operations have been considered in detail, showing both the underlying theoretical techniques and the corresponding experimental schemes. We stress that the method works when some a priori knowledge on the quantum channel is available, it requires fewer measurements than quantum process tomography and it is achievable experimentally with present day technology [12]. In particular, the detection method for non entanglement breaking channels and for non separable maps has already been demonstrated in a quantum optical experiment [13].

Acknowledgements

We thank Shashank Virmani for fruitful suggestions.

References

[1] Macchiavello C and Rossi M 2013 *Phys. Rev. A* **88** 042335
[2] Macchiavello C and Rossi M 2013 *Phys. Scripta* **T153** 014044
[3] Peres A 1996 *Phys. Rev. Lett.* **77** 1413; Horodecki P 1997 *Phys. Lett. A* **232** 333
[4] Virmani S, Huelga S F and Plenio M B 2005 *Phys. Rev. A* **71** 042328
[5] Horodecki M, Horodecki P and Horodecki R 1996 *Phys. Lett. A* **223** 1; Terhal B M 2000 *Phys. Lett. A* **271** 319

[6] Jamiolkowski A 1972 *Rep. Math. Phys.* **3** 275; Choi M D 1975 *Linear Algebr. Appl.* **10** 285
[7] Horodecki M, Shor P W and Ruskai M B 2003 *Rev. Math. Phys.* **15** 629
[8] Gühne O, Hyllus P, BrußD, Ekert A, Lewenstein M, Macchiavello C and Sanpera A 2002 *Phys. Rev. A* **66** 062305
[9] Gühne O, Hyllus P. BrußD, Ekert A, Lewenstein M, Macchiavello C and Sanpera A 2003 *J Mod. Opt.* **50** 1079
[10] Cirac J I, Dür W, Kraus B and Lewenstein M 2001 *Phys. Rev. Lett.* **86** 544
[11] Nielsen M A and Chuang I L 2000 *Quantum computation and quantum information* (Cambridge University Press)
[12] See for example: Chiuri A*et al.* 2010 *Phys. Rev. Lett.* **105** 250501; Chiuri A*et al.* 2011 *Phys. Rev. Lett.* **107** 253602
[13] Orieux A, Sansoni L, Persechino M, Mataloni P, Rossi M and Macchiavello C 2013 *quant-ph/1307.6005*

Journal of Physics: Conference Series **470** (2013) 012006 doi:10.1088/1742-6596/470/1/012006

Spherical Collapse in the Symmetron Model

Laura Taddei[1]

[1]Physics Department, University of Parma, Viale G.P. Usberti 7/A, 43100 Parma, Italy

E-mail: laura.taddei@fis.unipr.it

Abstract. The Symmetron model is a scalar field model for dark energy in which the scalar field is coupled to matter and is invariant under a reflection symmetry $\phi \to -\phi$. The coupling of the symmetron to matter is such that the symmetry is exact in high density environments and spontaneously broken when the density drops below a critical value. We study the non-linear evolution of density perturbations using the spherical collapse model. We calculate the extrapolated linear density contrast δ_c which is a fundamental quantity to obtain the halo-mass function for this model.

1. Introduction

In recent years, there has been much interest in exploring the possibility that the cosmic acceleration of the universe could be caused by a scalar field rather than by a cosmological constant. In the first class of models (called *quintessence models*), the scalar fields generally mediate a long range force of gravitational strength, which would induce dramatic violations of the equivalence principle, strongly constrained by local tests of gravity. Over the last decade, however, it has been realized that these scalar fields can mediate a long range force ($\sim$ Mpc) while satisfying local tests of gravity by imposing a *screening mechanism*, such as the Chameleon Mechanism [1], the Vainshtein Mechanism [2] and the Symmetron Mechanism, that we will discuss below.

2. The Symmetron model

Scalar-tensor theories are characterized by a conformal factor and by their interaction potential. In such a context, the symmetron model was proposed in [3, 4, 5] and is described by this action:

$$S = \int d^4x\sqrt{-g}\left\{\frac{M_{Pl}^2}{2}\mathcal{R} - \frac{1}{2}(\partial\phi)^2 - V(\phi)\right\} + \int d^4x\sqrt{-\tilde{g}}\;\; \mathcal{L}_m(\psi_i, \tilde{g}_{\mu\nu}) \tag{1}$$

where g is the determinant of the metric $g_{\mu\nu}$, R is the Ricci scalar, ψ_i are the different matter fields and $M_{Pl} \equiv \frac{1}{\sqrt{8\pi G}}$ where G is Newton's constant. Matter fields are minimally coupled to the Jordan frame metric $\tilde{g}_{\mu\nu}$, conformally related to the Einstein frame metric $g_{\mu\nu}$ by:

$$\tilde{g}_{\mu\nu} \equiv A^2(\phi)\, g_{\mu\nu} \tag{2}$$

Varying the action with respect to the scalar field ϕ, we obtain the field equations for ϕ in the Einstein frame [1]

$$\Box\phi = \frac{\partial V(\phi)}{\partial\phi} + \rho\frac{\partial A(\phi)}{\partial\phi} \tag{3}$$

So the field evolves according to an effective potential

$$V_{eff}(\phi) = V(\phi) + \rho A(\phi) \tag{4}$$

For the symmetron model of interest, we choose:

$$V(\phi) = \bar{V} + V_0 e^{-\frac{\phi^2}{2M^2}} \tag{5}$$

and

$$A(\phi) = e^{\frac{\lambda\phi^2}{2M^2}} \tag{6}$$

in which $\bar{V}$ is the cosmological constant energy density, V_0 is a small energy density fixed by the phase transition, λ is one dimensionless coupling constant and M is the mass scale. So we have:

$$V_{eff}(\phi) = \bar{V} + V_0 e^{-\frac{\phi^2}{2M^2}} + \rho e^{\frac{\lambda\phi^2}{2M^2}} \tag{7}$$

Since $\phi^2 \ll M^2$, we can expand $V_{eff}(\phi)$ in Taylor series so we obtain the effective potential described in [5]. With the above choice for the signs (we will assume $\lambda > 0$), there is a density-dependent phase transition. Indeed, looking at the second derivative of the potential in $\phi = 0$,

$$M^2 \frac{d^2 V_{eff}}{d\phi^2}|_{\phi=0} = -V_0 + \lambda\rho \tag{8}$$

we see that it changes sign at

$$\rho(z_t) = \rho_0 (1 + z_t)^3 = \frac{V_0}{\lambda} \tag{9}$$

where z_t is the redshift in which the phase transition occurs.
As long as ρ is high enough ($z \geq z_t$), the minimum of V_{eff} exhibits the $\mathbb{Z}_2$ symmetry and the vacuum expectation value (VEV) goes to zero ($\phi_{VEV} = 0$). In low density regions ($z \leq z_t$), instead, the potential breaks the reflection symmetry spontaneously and the scalar acquires a VEV:

$$\phi_{VEV} = \phi_0(z) = \pm M\sqrt{\frac{6}{\lambda+1}\log\left(\frac{1+z_t}{1+z}\right)} \tag{10}$$

where we have chosen the minimum with the " + " sign. Fluctuations $\delta\phi$ around a local background value ϕ_{VEV}, as would be detected by local experiments, couple as $\approx \frac{\lambda}{M^2}\phi_{VEV}\delta\phi\rho$, that is, the coupling is proportional to the local VEV.

2.1. Scalar coupling

We define the coupling of the scalar field to matter $\beta(\phi)$ as in [6]:

$$\beta(\phi = \phi_0) = \beta_0(z) = M_{Pl}\frac{d\log A(\phi)}{d\phi}|_{\phi=\phi_0} = M_{Pl}\frac{\lambda\phi_0(z)}{M^2} \tag{11}$$

[1] where $\rho = A^3\tilde{\rho}$ is the matter density, which is conserved in the Einstein frame and $\tilde{\rho}$ is the 00 component of the energy-momentum tensor $\tilde{T}_{\mu\nu} = -\left(2/\sqrt{-\tilde{g}}\right)\delta\mathcal{L}_m/\delta\tilde{g}^{\mu\nu}$ in the Jordan frame, which is covariantly conserved: $\tilde{\nabla}_\mu \tilde{T}^\mu_\nu = 0$.

where $\phi_0 = 0$ before the phase transition, and $\phi_0 \neq 0$ after the phase transition.
We take $\beta_0 (z = 0)$ as a parameter of our model:

$$\beta_0 (z = 0) = \frac{\lambda M_{Pl}}{M} \sqrt{\frac{6}{\lambda + 1} \log (1 + z_t)} \cong \frac{\lambda M_{Pl}}{M} \sqrt{6 \log (1 + z_t)} \tag{12}$$

because $\lambda \ll 1$. We have fixed $\beta_0 \cong 0.75$ from the value of the error on the parameter σ_8 [2] which is about 5% and the phase transition $z_t = 1$ near today.

2.2. *Spherical solutions*

To study the implication for tests of gravity, we follow the same steps of [5, 6]. In analogy with the Chameleon Mechanism (see [1]), we define a dimensionless parameter γ:

$$\gamma \equiv \frac{\lambda}{M^2} (\rho - \bar{\rho}) R^2 = 6 \frac{M_{Pl}^2}{M^2} \Phi \tag{13}$$

Physically, this ratio measures the surface Newtonian potential Φ relative to M/M_{Pl}.
We define *screened objects* the case in which the scalar force is confined only within a thin-shell beneath the surface and *unscreened objects* the case in which the scalar force is non zero inside the object. To satisfy experimental constraints we want the Milky Way to be screened so we set:

$$\frac{M}{M_{Pl}} = \mu \cong 10^{-3} \tag{14}$$

following [5].

3. Spherical collapse

A popular tool to study the non-linear growth in cold dark matter is the spherical collapse [7, 8]. This non-linear approximation was first used in the standard cold dark matter scenario, but later also in the cold dark matter model with a cosmological constant (ΛCDM) [9].

3.1. *Application to Symmetron model*

Now we study how to apply the spherical collapse to symmetron model.
At the beginning, the initial density contrast of the bubble is very small ($\delta_{m,i} = 0.0003$) so we start by $\gamma \ll 1$: initially the sphere is in the *thick-shell* regime and the scalar field has this value both inside and outside the sphere. Then, the transition from *thick-shell* to *thin-shell* regime depends on the initial radius of the bubble (R_i). For small values of the initial radius, the sphere remains in the *thick-shell* regime for all the time that it takes to collapse and we have that the scalar field is inside the bubble and in the background; in this case, the γ parameter is always less than one. For large initial radii, the sphere is initially in the *thick-shell* regime; at some point, the γ parameter becomes larger than one and we turn off the scalar field inside the sphere. In this case we observe the transition from *thick* to *thin-shell* regime. We denote by ϕ the scalar field inside the bubble and by $\bar{\phi}$ the scalar field in the background.
The flat background Universe is described by Friedmann equations:

$$\left(\frac{\ddot{a}}{a}\right) = -\frac{1}{6M_{Pl}^2} \left(\bar{\rho}_m A\left(\bar{\phi}\right) + \bar{\rho}_\phi + 3\bar{p}_\phi\right) \tag{15}$$

where:

$$\bar{\rho}_\phi = \frac{1}{2}\dot{\bar{\phi}}^2 + V\left(\bar{\phi}\right) \tag{16}$$

[2] where σ_8 parameter is defined as: $\sigma_8^2\left(r = 8Mpc/h\right) = \int d^3k/\left(2\pi\right)^3 |\tilde{W}\left(kr\right)|^2 P_L\left(k\right)$.

$$\bar{p}_\phi = \frac{1}{2}\dot{\bar{\phi}}^2 - V\left(\bar{\phi}\right) \tag{17}$$

are respectively the density and the pressure of the scalar field and

$$\bar{H}^2 \equiv \left(\frac{\dot{a}}{a}\right)^2 = \frac{1}{3M_{Pl}^2}\left(\bar{\rho}_m A\left(\bar{\phi}\right) + \bar{\rho}_\phi\right) \tag{18}$$

We can derive the evolution equation for the bubble radius R by following the same steps of [10]. So we have:

$$\left(\frac{\ddot{R}}{R}\right) = -\beta\left(\phi\right)\dot{\bar{\phi}}\left(\bar{H} - \frac{\dot{R}}{R}\right) + \left(\frac{\ddot{a}}{a}\right) - \frac{1}{6M_{Pl}^2}\bar{\rho}_m\delta_m\left(1 + 2\beta\left(\phi\right)^2\right) \tag{19}$$

Since the scalar field is rolling slowly along the minimum, we can neglect all terms proportional to $\dot{\bar{\phi}}$. In the *thick-shell* regime the scalar field is inside the sphere and in the background: $\phi = \bar{\phi} = \phi_0$. So we use equations 15-19 in which we put $\phi = \phi_0$.
In the transition regime from *thick* to *thin-shell*, at the beginning we use the equations of the previous case. When γ parameter becomes bigger than one, we put $\phi_0 = 0$. The equations are the same for the background, while the evolution equations for the radius R become:

$$\left(\frac{\ddot{R}}{R}\right) = \left(\frac{\ddot{a}}{a}\right) - \frac{1}{6M_{Pl}^2}\bar{\rho}_m\delta_m \tag{20}$$

when we turn-off the scalar field inside the sphere. The important thing is that the dynamics

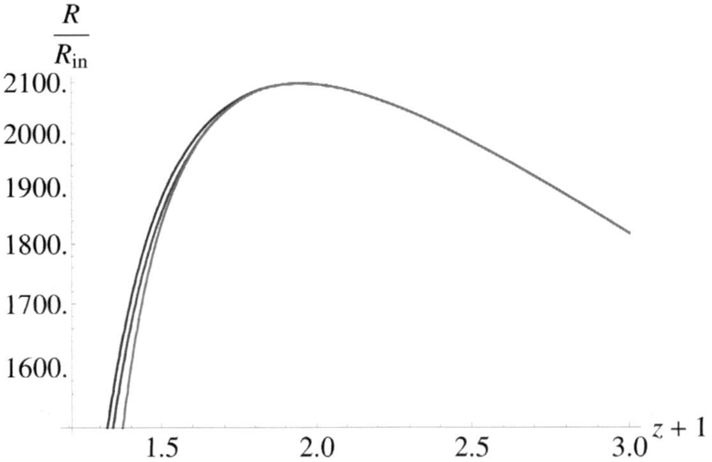

Figure 1. Evolution of radius $\frac{R}{R_i}$ vs $(z+1)$ for $\delta_{m,i} = 0.0003$. The blue curve represents ΛCDM model, the red curve represents the transition regime from *thick* to *thin-shell* obtained by fixing $R_i H_0 = 0.45$ and the yellow curve represents the *thick-shell* regime, obtained by fixing $R_i H_0 = 0.1$.

of collapse and the type of regime depends on the initial radius of the sphere (R_i). For small initial radius ($R_i H_0 \leq 0.1$ means $R_i \leq 300$ Mpc), we have that $\gamma < 1$ for all the time. If we integrate numerically equations 15-19 from $z_i \sim 7000$ to $z_f = 0$, with an initial density contrast $\delta_{m,i} = 0.0003$, we observe that the radius collapses at $z \cong 0.2$, as we can see from figure 1. For large initial radius ($0.1 < R_i H_0 \leq 0.45$ means R_i between about 300 and 1400 Mpc) we have the transition regime: we start by integrating numerically the same equations of the previous case from $z_i \sim 7000$ to the moment in which $\gamma \geq 1$ ($z \sim 0.7$). Then we use equation 20 for radius evolution when we pass to *thin-shell* regime until the end $z_f = 0$. In this case, the scalar force is confined only within a *thin-shell* beneath the surface so the sphere collapses later respect to the case in which the object is unscreened.

3.2. Determination of δ_c

We follow the same steps of [10] to obtain the linear evolution of density contrast:

$$\ddot{\delta}_{m,L} = \left(\beta\dot{\bar{\phi}} - 2\bar{H}\right)\dot{\delta}_{m,L} + \frac{1}{a^2}\nabla^2\Phi_{eff} \qquad (21)$$

where Φ_{eff} is the effective gravitational potential given by

$$\Phi_{eff} \equiv \Phi + \beta\delta\phi \qquad (22)$$

which follows the modified Poisson equation

$$\nabla^2\Phi_{eff} = -\frac{a^2}{2M_{Pl}^2}\bar{\rho}_m\delta_m\left(1 + 2\beta^2\right) \qquad (23)$$

If we assume the adiabatic approximation, we can neglect all terms proportional to $\dot{\bar{\phi}}$ so equation 21 becomes:

$$\ddot{\delta}_{m,L} \cong -2\bar{H}\dot{\delta}_{m,L} + \frac{1}{a^2}\nabla^2\Phi_{eff} \qquad (24)$$

where $\bar{H}$ is defined in equation 18. We numerically solve equation 24 from $z \sim 7000$ to $z = 0$ and we calculate the value of δ at redshift of collapse z_c by varying the value of $\delta_{m,i}$. From figure 2, the two models (*thick-shell* regime and transition from *thick* to *thin-shell* regime) approach ΛCDM at high redshift after the phase transition. At $z = 0$, the difference in δ_c with respect to ΛCDM is about 2% for *thin-shell* regime, while is about 4% for *thick-shell* regime.

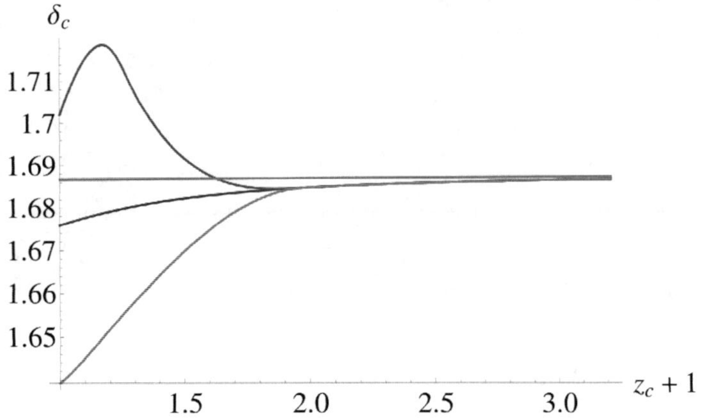

Figure 2. Extrapolated linear density contrast at collapse δ_c vs. $z_c + 1$ for Einstein de-Sitter EdS (green curve), ΛCDM (blue curve), transition from *thick* to *thin-shell* regime (red curve) and *thick-shell* regime (yellow curve) cases.

4. Halo-mass function

We use a model of halo-mass function based on spherical collapse and on the Sheth-Tormen (ST) prescription, by following the procedure of [11]. We calculate the halo-mass function by fixing the value of δ_c at $z = 0$, as we can see from figure 3.

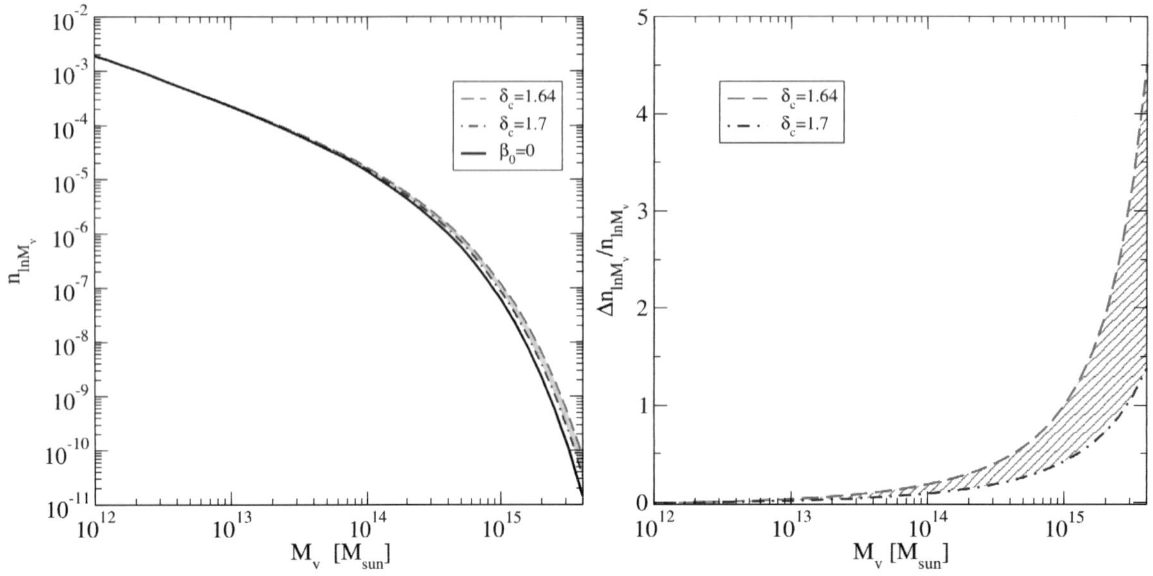

Figure 3. Left panel: Halo mass function as a function of the virialization mass M_v for *thick-shell* regime (red dashed line) by fixing $\delta_c(z=0) \simeq 1.64$ and $\beta_0 = 0.75$, for the transition regime (blue dot-dashed line) by fixing $\delta_c(z=0) \simeq 1.70$ and $\beta_0 = 0.75$. The green band shows the region between the two values of δ_c. The solid black line corresponds to the ΛCDM case, i.e. $\beta_0 = 0$. Right panel: Relative deviations of the halo mass function from ΛCDM model in function of M_v.

5. Conclusions

Dark matter haloes are the building blocks of cosmological observables associated with structures in the universe. In this model, the scalar coupling which occurs after the phase transition leads to an enhanced abundance of massive halos with respect to the Standard Model of Cosmology. In figure 3 (right panel) we observe that the number of haloes increases significantly, especially at the high mass end, by up to 50-100% for cluster-sized haloes. The transition regime (blue curve) suppresses the abundance of haloes in the high mass end, as expected. A similar behavior was found in [11] for $f(R)$ models.

6. Acknowledgments

I would like to thank my supervisor Massimo Pietroni for helping me in carrying on the work and Riccardo Catena for doing the plots of the last part of this article.

7. References

[1] Khoury J and Weltman A 2004 *Phys. Rev. D* **69** 044026
[2] Vainshtein A I 1972 *Phys. Lett. B* **39** 393
[3] Olive K A and Pospelov M 2008 *Phys. Rev. D* **77** 043524
[4] Pietroni M 2005 *Phys. Rev. D* **72** 043535
[5] Hinterbichler K and Khoury J 2010 *Phys. Rev. Lett.* **104** 231301
[6] Hinterbichler K, Khoury J, Levy A and Matas A 2011 *Phys. Rev. D* **84** 103521
[7] Padmanabhan T 1995 *Structure formation in the Universe* (Cambridge University Press)
[8] Peacock J 1999 *Cosmological Physics* (Cambridge University Press)
[9] Lahav O, Lilje P B, Primack J R and Rees M J 1991 *Mon. Not. Roy. Astron. Soc.* **251** 128
[10] Wintergerst N and Pettorino V 2010 *Phys. Rev. D* **82** 103516
[11] Schmidt F, Lima M V, Oyaizu H and Hu W 2009 *Phys. Rev. D* **79** 083518

Journal of Physics: Conference Series **470** (2013) 012007 doi:10.1088/1742-6596/470/1/012007

Cosmological constraints on variations of the fine structure constant at the epoch of recombination

E Menegoni[1], S Galli[2,3], M Archidiacono[4], E Calabrese[5] and A Melchiorri[6]

[1]The Institut für Theoretische Physik, University of Heidelberg, Philosophenweg 16, 69120 Heidelberg, Germany.
[2]Institut d'Astrophysique de Paris, UMR-7095 du CNRS, Universite' Pierre et Marie Curie, 98 bis bd Arago, 75014 Paris, France.
[3] Sorbonne Universite's, Institut Lagrange de Paris, 98 bis bd Arago, 75014 Paris, France
[4] Department of Physics and Astronomy University of Aarhus, DK-8000 Aarhus C, Denmark.
[5] Sub-department of Astrophysics, University of Oxford, Denys Wilkinson Building, Keble Road, Oxford, OX1 3RH, UK.
[6] Dipartimento di Fisica, Universita' La Sapienza, P. le A. Moro 2, 00185 Roma, Italy.

E-mail: e.menegoni@thphys.uni-heidelberg.de

Abstract. In this brief work we investigate any possible variation of the fine structure constant at the epoch of recombination. The recent measurements of the Cosmic Microwave Background anisotropies at arcminute angular scales performed by the ACT and SPT experiments are probing the damping regime of Cosmic Microwave Background fluctuations. We study the role of a mechanism that could affect the shape of the Cosmic Microwave Background angular fluctuations at those scales, namely a change in the recombination process through variations in the fine structure constant α.

1. Introduction

The recent observations from Cosmic Microwave Background (CMB hereafter) satellite, balloon-borne and ground based experiments ([1]-[4]), galaxy redshift surveys [5] and luminosity distance measurements, have fully confirmed the theoretical predictions of the standard ΛCDM cosmological model. This not only allows to place stringent constraints on the parameters of the model but can be fruitfully used to constrain non standard physics at the fundamental level. The ΛCDM model assumes the validity of General Relativity on cosmological scales, as well as the physics of the standard model of particle physics. One possible extension, which may have motivations in fundamental physics, is to consider variations of dimensionless constants [6, 7]. Possible changes in the recombination process have been investigated by several authors. Another possible mechanism is based on the hypothesis of a change in the fundamental constants of nature, specifically the fine structure constant, α ([8]-[11]).

In this framework, an interesting discrepancy with the expectations of the standard model has recently been discovered in the small CMB scale measurements of the ACT [2] and SPT [4] experiments. The effective number of relativistic degrees of freedom N_{eff} (for a rewiev on N_{eff} see [12]) has been reported as higher (at more than two standard deviations) than the expected standard value of $N_{eff} = 3.046$ (the little deviation from $N_{eff} = 3$ takes into account effects

Journal of Physics: Conference Series **470** (2013) 012007 doi:10.1088/1742-6596/470/1/012007

from the non-instantaneous neutrino decoupling from the primordial photon-baryon plasma, see for more deatails [13]). This discrepancy has been confirmed by several recent analyses of the ACT and SPT datasets (see e.g. [14]-[20]). We analyse this issue by obtaining constraints in the framework of a non-standard recombination process, in which variations of the fine structure constant are permitted.

2. Method and Analysis

The analysis method we adopted is based on the publicly available Monte Carlo Markov Chain package `cosmomc` [21] with a convergence diagnostic done through the Gelman and Rubin statistic. A COSMOMC analysis [21] was performed combining the following CMB datasets: WMAP7 [1], ACT [2], ACBAR [3], SPT [4]: we analyzed datasets out to $l_{\rm max} = 3000$, and we include also information on dark matter clustering by using the SDSS-DR7 luminous red galaxy sample [5]. We impose a prior on the Hubble parameter based on the last Hubble Space Telescope observations [22].

We sample the following six-dimensional standard cosmological parameters, adopting flat priors on them: the baryon and cold dark matter densities $\Omega_b h^2$ and $\Omega_c h^2$, the Hubble constant H_0, the optical depth to reionization τ, the scalar spectral index n_S, and the overall normalization of the spectrum A_S [23]. In order to perform this analysis, we consider purely adiabatic initial conditions and we impose spatial flatness. The variations in the fine structure constant α/α_0 where α_0 is the current, local, value are implemented in the code by modifying the RECFAST recombination subroutine ([24] -[27]) following the procedure described in [28].

Table 1. In the table are reported the results for the three analyses described in the text by using a MCMC estimation of the cosmological parameters from the dataset described in the Analysis section. Upper bounds at 95% c.l. are reported for foregrounds parameters [23].

Parameter	α/α_0	α/α_0+N_{eff}	α/α_0+N_{eff}+Y_p
$\Omega_b h^2$	0.0218 ± 0.0004	0.0224 ± 0.0005	0.0223 ± 0.0007
$\Omega_c h^2$	0.1144 ± 0.0034	0.1302 ± 0.0095	0.1303 ± 0.0094
τ	0.086 ± 0.014	0.088 ± 0.015	0.088 ± 0.016
H_0	68.9 ± 1.4	71.52 ± 2.0	71.8 ± 2.1
α/α_0	0.984 ± 0.005	0.99 ± 0.006	0.987 ± 0.014
n_s	0.976 ± 0.013	0.991 ± 0.015	0.992 ± 0.016
$log[10^{10} A_s]$	3.193 ± 0.037	3.169 ± 0.04	3.167 ± 0.042
A_{SZ}	$< 2.$	$< 2.$	$< 2.$
A_C	$< 16.$	< 15.8	< 14.8
A_P	< 24.7	< 24.9	< 22.4
Ω_Λ	0.7137 ± 0.007	0.702 ± 0.0094	0.704 ± 0.013
Age/Gyr	13.76 ± 0.24	13.18 ± 0.38	13.15 ± 0.37
Ω_m	0.2863 ± 0.007	0.298 ± 0.0094	0.296 ± 0.013
σ_8	0.836 ± 0.023	0.862 ± 0.028	0.859 ± 0.034
z_{re}	10.7 ± 1.2	$11. \pm 1.3$	$11. \pm 1.3$
N_{eff}	–	$4.10^{+0.24}_{-0.29}$	$4.19^{+0.31}_{-0.35}$
Y_p	–	–	0.215 ± 0.096
χ^2_{min}	7600.2	7596.8	7596.5

Also variations in the effective number of relativistic degrees of freedom N_{eff} and the primordial Helium abundance Y_p, otherwise fixed at the values $N_{eff} = 3.046$ and $Y_p = 0.24$, are included. Since we are varying also the Helium abundance, we considered variations in α also in the process of Helium recombination. A $\sim 5\%$ change of α for Helium recombination changes the CMB angular spectra by less than 0.5% up to $l = 1500$. We focus on constraing the variations of the fine structure constant at the epoch of Recombination and we fixed the fine structure constant to the local standard value $\alpha = \alpha_0$ in the epoch of reionization [23].

The dataset considered prefers a value of α/α_0 smaller than unity at more than two standard deviations when both the N_{eff} and Y_p are kept fixed at their standard values as shown in the table 1. This result was to be expected since is clearly driven by data preference for larger values of N_{eff}. Allowing for variations in N_{eff} significantly shifts the best fit value for α/α_0, causing it to be now consistent with the standard value.

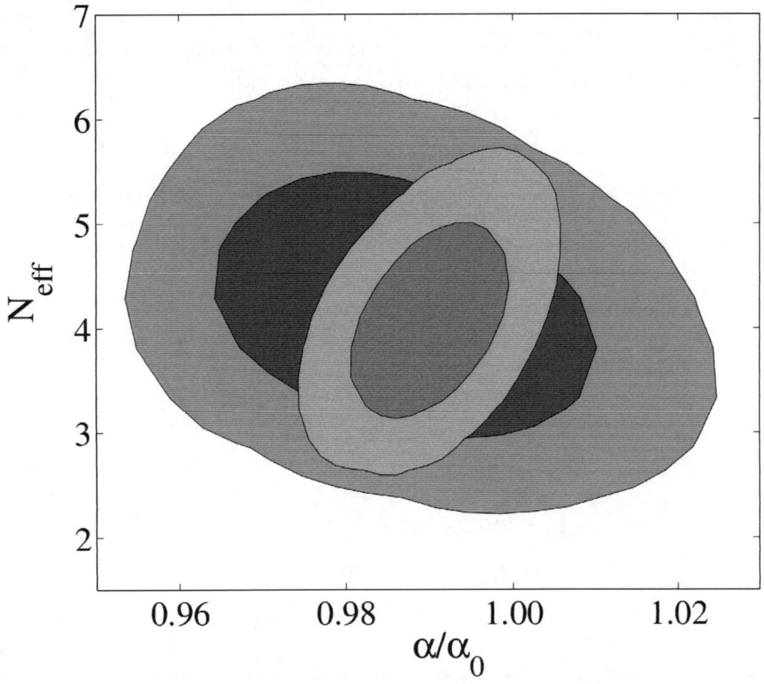

Figure 1. Likelihood contour plot for α/α_0 vs N_{eff} at 68% c.l. and 95% c.l. in the case of $Y_p = 0.24$ (red smaller contours) and Y_p allowed to vary (blue larger contours).

Allowing for variation in the fine structure constant enlarges the error bars on N_{eff} of about $\sim 30\%$ but does not shift the best fit value towards the standard result. The largest effect on α comes however when also the helium abundance Y_p is let free to vary: the errors on α are almost doubled [23].

We can better understand the impact of Y_p on the determination of α/α_0 by looking at figure 1, where we plot the 2-D likelihood contours in the α/α_0-N_{eff} plane in the cases of $Y_p = 0.24$ and free Y_p. In the case in which we keep the helium abundance fixed there is a clear but moderate degeneracy between α/α_0 and N_{eff}, this is because when N_{eff} is increased the Hubble parameter at recombination increases. By decreasing the free electron density at recombination we recover the damping scale at the same value fixed by observations [23].

Journal of Physics: Conference Series **470** (2013) 012007 doi:10.1088/1742-6596/470/1/012007

The degeneracy changes direction by varying the Helium abundance, in fact, a larger value for Y_p produces a large free electron fraction at recombination and a smaller value for N_{eff} is needed to keep the damping scale small. On the other hand a large value for Y_p needs large values for α. So now small values of N_{eff} are more compatible with observations when α is larger.

The other cosmological parameters are effected in different ways when we let vary the value of N_{eff} and Helium abundance Y_p: in particular the value of the "standard" matter is a little bit larger (both $\Omega_b h^2$ and $\Omega_c h^2$ see table [1]) respect to the case in which N_{eff} and Y_p are kept fixed, and the value of Ω_Λ is smaller (table [1]), furthermore, the recombination process happens at higher value of the redshift z_{re}, meaning that the formation of neutral hydrogen is accelerated respect to the standard case.

3. Conclusions

We presented new constraints on variation of the fine structure constant using the recent CMB anisotropy measurements from the ACT and SPT experiments, combined with other cosmological datasets. Assuming the standard three active neutrino framework and a primordial Helium abundance of $Y_p = 0.24$ the current data favours a lower value for the fine structure constant at more than two standard deviations with $\alpha/\alpha_0 = 0.984 \pm 0.005$. When the number of relativistic degrees of freedom is let to vary freely, the standard value is consistent with the data considered, while varying the primordial Helium abundance further enlarges the error bars.

The recent Planck data delivered on March 2013 shows an improvement on the constraints on the variations of the fine structure constant by a factor of about five if compared with those from WMAP-9 data. The analysis of Planck data limits any variation in the fine structure constant from the epoch of recombination to the present day to be less than approximately 0.4% (for more details see [29]).

References

[1] Komatsu E *et al.* 2011 *Astrophys. J. S.* **192** 18
[2] Dunkley J *et al.* 2011 *Astrophys. J.* **739** 52
[3] Reichardt C L *et al.* 2009 *Astrophys. J.* **694** 1200
[4] Keisler R *et al.* 2011 *Astrophys. J.* **743** 28
[5] Reid B A *et al.* 2010 Mon. Not. Roy. Astron. Soc. **404** 60
[6] Uzan J-P 2003 *Rev. Mod. Phys.* **75** 403
[7] Uzan, J-P 2011 *Living Rev. Rel.* **14** 2
[8] Stefanescu P 2007 *New Astron.* **12** 635
[9] Nakashima M, Nagata R and Yokoyama J. 2008 *Prog. Theor. Phys.* **120** 1207
[10] Scoccola C G, Landau S J, Vucetich H 2008 *Phys. Lett. B* **669** 212-216
[11] Landau S J, Scoccola C G 2010 *Astronomy and Astrophysics* **517** A62
[12] Kolb E W and Turner M S 1990 *Front. Phys.* **69** 1
[13] Mangano G, Miele G, Pastor S, Pinto T, Pisanti O, Serpico P D 2005 *Nucl. Phys. B* **729** 221-234
[14] Archidiacono M, Calabrese E and Melchiorri A 2011 *Phys. Rev. D* **84**, 123008
[15] Hou Z, Keisler R, Knox L, Millea M, Reichardt C 2013 *Phys. Rev. D* **87** 083008
[16] Smith T L, Das S and Zahn O 2012 *Phys. Rev. D* **85** 023001
[17] Hamann J *J. Cosmology Astropart. Phys.* **3** 21
[18] Giusarma E, Corsi M, Archidiacono M, de Putter R, Melchiorri A, Mena O, Pandolfi S 2011 *Phys. Rev. D* **83** 115023
[19] Giusarma E, Archidiacono M, de Putter R, Melchiorri A, Mena O 2012 *Phys. Rev. D* **85** 083522
[20] Smith A, Archidiacono M, Cooray A, De Bernardis F, Melchiorri A and Smidt J 2012 *Phys. Rev. D* **85** 123521
[21] Lewis A and Bridle S 2002 *Phys. Rev. D* **66** 103511 (Available from `http://cosmologist.info`.)
[22] Riess A G *et al.* 2011 *Astrophys. J.* **730** 119
[23] Menegoni E, Archidiacono M, Calabrese E, Galli S, Martins C J A P, Melchiorri A 2012 *Phys. Rev. D* **85** 107301

[24] Avelino P P *et al.* 2001 *Phys. Rev. D* **64** 103505; Martins C J A, Melchiorri A, Trotta R, Bean R, Rocha G, Avelino P P and Viana P T P 2002 *Phys. Rev. D* **66** 023505
[25] Martins C J A, Melchiorri A, Rocha G, Trotta R, Avelino P P and Viana, P T P 2004 *Phys. Lett. B* **585** 29
[26] Rocha G, Trotta R, Martins C J A, Melchiorri A, Avelino P P, Bean R and Viana P T P 2004 *Mon. Not. Roy. Astron. Soc.* **352** 20
[27] Hannestad S 1999 *Phys. Rev. D* **60** 023515
[28] Menegoni E, Galli S, Bartlett J G, Martins C J A and Melchiorri A 2009 *Phys. Rev. D* **80** 087302; Menegoni E et al. 2010 *Intern. J. Mod. Phys. D* **19** 04 507-512; Calabrese E, Menegoni E, Martins C J A P, Melchiorri A and Rocha G 2011 *Phys. Rev. D* **84** 023518
[29] Planck Collaboration 2013 Planck 2013 results. XVI. Cosmological parameters *Preprint* astro-ph/1303.5076

Journal of Physics: Conference Series **470** (2013) 012008 doi:10.1088/1742-6596/470/1/012008

Bar-mode instability suppression in magnetized relativistic stars

L Franci[1], R De Pietri[1], K Dionysopoulou[2] and L Rezzolla[2,3]

[1] Dipartimento di Fisica e Scienze della Terra, Università di Parma and INFN, Parma, Italy
[2] Max-Planck-Institut für Gravitationsphysik, Albert-Einstein-Institut, Golm, Germany
[3] Institut für Theoretische Physik, Frankfurt am Main, Germany

E-mail: luca.franci@fis.unipr.it

Abstract. We show that magnetic fields stronger than about 10^{15} G are able to suppress the development of the hydrodynamical bar-mode instability in relativistic stars. The suppression is due to a change in the rest-mass density and angular velocity profiles due to the formation and to the linear growth of a toroidal component that rapidly overcomes the original poloidal one, leading to an amplification of the total magnetic energy. The study is carried out performing three-dimensional ideal-magnetohydrodynamics simulations in full general relativity, superimposing to the initial (matter) equilibrium configurations a purely poloidal magnetic field in the range $10^{14} - 10^{16}$ G. When the seed field is a few parts in 10^{15} G or above, all the evolved models show the formation of a low-density envelope surrounding the star. For much weaker fields, no effect on the matter evolution is observed, while magnetic fields which are just below the suppression threshold are observed to slow down the growth-rate of the instability.

1. Introduction

Differentially rotating neutron stars (NSs) are subject to the so-called $m = 2$ dynamical bar-mode instability for non-radial axial modes with azimuthal dependence $e^{im\phi}$ ($m = 1, 2, ...$), when the instability parameter $\beta \equiv T/|W|$ (i.e. the ratio between the rotational kinetic energy T and the gravitational binding energy W) exceeds a critical value. The bar-mode instability in differentially rotating magnetized NSs has already been studied in the Newtonian case by Camarda et al. [1], suggesting that the effect of magnetic fields on the emergence of the instability is not likely to be very signicant unless when NSs are born very highly magnetized.

In this work, we investigate if the presence of magnetic fields can affect the onset and development of this kind of instability in full general relativity, as well as the role played by the magnetic braking to possibly suppress the instability.

Our study is motivated by the potential implications that strong magnetic fields may have for astrophysical scenarios and for gravitational wave astronomy. In particular, during the last decade a general consensus has formed that long Gamma-Ray Bursts (GRBs) arise from the collapse of massive stars while short GRBs, most likely, from neutron star mergers. A key ingredient for both these categories of astrophysical phenomena is the formation of magnetic structures that could power the high-speed particle jets associated with GRBs [2].

We choose to evolve different equilibrium relativistic stellar models that have already been studied in the non-magnetized case by Baiotti et al. [3], so that their behavior against the bar-mode instability is already known and fully understood when no magnetic fields are present.

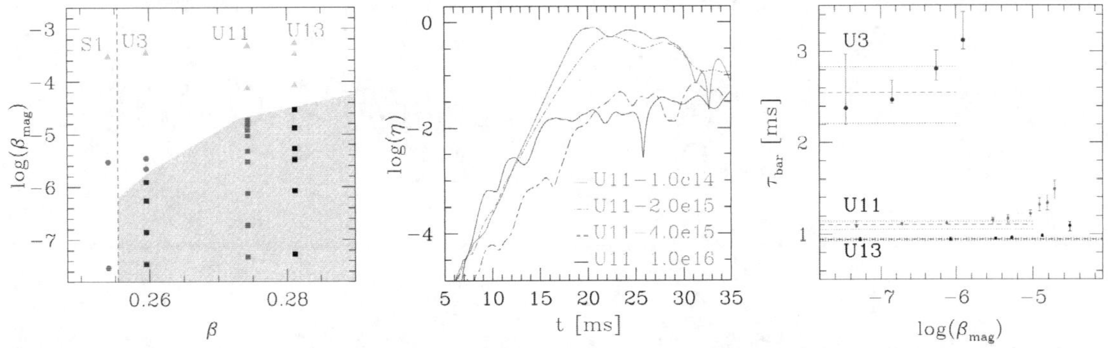

Figure 1. (Left panel): parameter space (β, $\beta_{\rm mag}$) with the initial values of these two quantities for all the simulated models: the models inside the red shaded region turned out to be still matter-unstable even when a magnetic field is present. (Central panel): time evolution of the distortion parameter η for model `U11` for four different values of the seed poloidal magnetic field strength. We use blue lines for matter-unstable models and black lines for the stable ones. (Right panel): growth-time τ_{bar} of the bar-mode instability versus the initial values of $\beta_{\rm mag}$. The dashed lines represent the values in the non-magnetized cases while the dotted lines stand as error bars.

2. Initial data

The initial data for our simulations are computed as stationary equilibrium solutions of axisymmetric relativistic stars that are rapidly differentially rotating. In particular, we assume the non-uniform j-law angular velocity distribution

$$\Omega_c - \Omega = \frac{1}{\hat{A}^2 R_e^2}\left[\frac{(\Omega-\omega)r^2\sin^2\theta e^{-2\nu}}{1-(\Omega-\omega)^2r^2\sin^2\theta e^{-2\nu}}\right], \tag{1}$$

where R_e is the coordinate equatorial stellar radius, the coefficient $\hat{A}$ is a measure of the degree of differential rotation, while ω and ν are metric components in spherical coordinates.

A seed poloidal magnetic field is added to these initial equilibrium models as a perturbation by introducing a purely toroidal vector potential $A_\phi = A_b\,(\max(p - p_{\rm cut}, 0))^2$, where $p_{\rm cut}$ is 4% of the maximum initial pressure ($p_{\rm cut} = 0.04\ \max(p(0))$) and A_b is chosen in such a way so that it corresponds to the maximum of the initial magnetic field on the (x, y) plane $B^z_{\rm max}|_{t,z=0}$.

The equilibrium models considered here have been calculated using a relativistic polytropic equation of state $p = K\rho^\Gamma$ with $K = 100$ and $\Gamma = 2$, in analogy with previous works in the literature. In particular, to better find out the changes to the bar-mode dynamics induced by the presence of a magnetic field, we focused our attention on a sequence of models having a fixed constant amount of differential rotation $\hat{A} = 1$ and a constant rest mass $M_0 \simeq 1.5\ M_\odot$, which were already studied in the non-magnetized case in [3]. All the features of these models are then described in further detail in table I of [3] and here we use the notation therein reported. For the present study, we selected the three models which were better discussed therein, namely models `U3`, `U11` and `U13`, where `U3` is the closest to the threshold for the onset of the bar-mode instability and `U13` is almost the fastest rotating equilibrium model which can be obtained for this sequence at constant barionic mass (see the left panel of figure 1, where the threshold is indicated with a red dotted line corresponding to $\beta = 0.255$).

Indeed, all the simulated initial models are represented in the left panel of figure 1, where they are identified in terms of the initial values of the two parameters (β, $\beta_{\rm mag}$), defined as $\beta \equiv T/|W|$ and $\beta_{\rm mag} \equiv E_{\rm mag}/(T + |W|)$, respectively. Here, T is the rotational kinetic energy, W the gravitational binding energy and $E_{\rm mag}$ the total magnetic energy. Hereafter, we will refer

Journal of Physics: Conference Series **470** (2013) 012008 doi:10.1088/1742-6596/470/1/012008

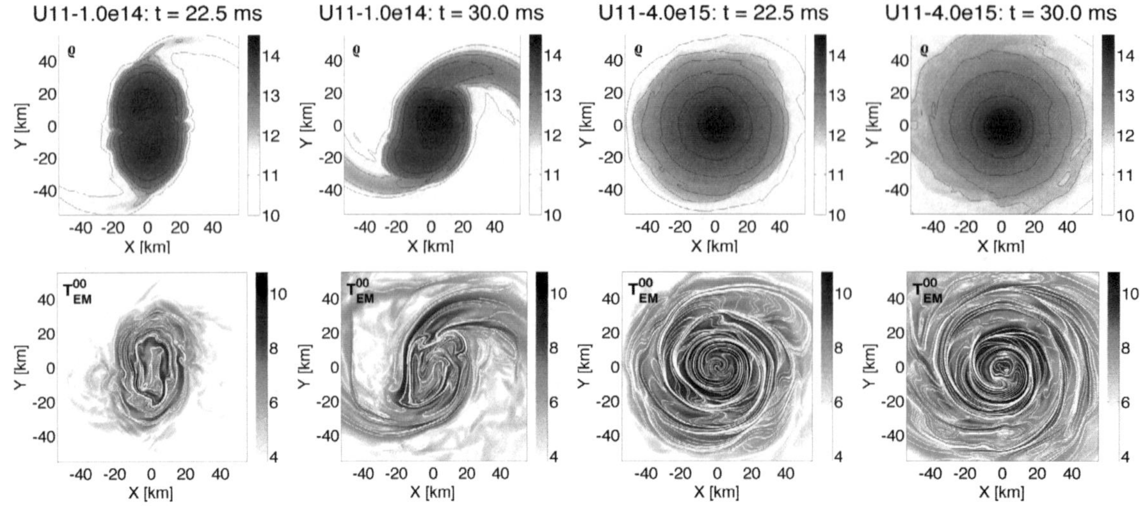

Figure 2. (Top panels): snapshots of the rest-mass density ρ on the $(x,\, y)$ plane at $t = 22.5$ and 30.0 ms for model `U11` with two different seed magnetic field strengths: $B^z_{\rm max}|_{t,z=0} = 1.0 \times 10^{14}$ G for model `U11-1.0e14` (left panels) and $B^z_{\rm max}|_{t,z=0} = 4.0 \times 10^{15}$ G for `U11-4.0e15` (right panels). The color code is defined in terms of $\log_{10}(\rho)$ where ρ is in cgs units (g/cm^3). Additionally, isodensity contours are shown for $\rho = 10^{11}$, 10^{12}, 5×10^{12}, 10^{13}, 5×10^{13} and 10^{14} g/cm^3; (Bottom panels): snapshots of the local magnetic energy $T^{00}_{\rm EM}$ on a horizontal plane at $z \simeq 1.5$ km for the same two models at the same steps during the evolution. Additionally, magnetic field lines are drawn in white. The color code is defined in terms of $\log_{10}(T^{00}_{\rm EM}/c^2)$ where $T^{00}_{\rm EM}/c^2$ is in cgs units (g/cm^3).

to a particular evolved stellar model by using the notation Uxx-yy, where xx is the initial model and yy denotes the maximum initial magnetic field strength (`U11-1.0e14` refers to a `U11` model with a super-imposed purely poloidal magnetic field such that $B^z_{\rm max}|_{t,z=0} = 1.0 \times 10^{14}$ G).

3. Computational setup

The simulations reported here have been performed using the general-relativistic magnetohydrodynamics (GRMHD) code WhiskyMHD described in [5] and which is based on the `Cactus` computational toolkit. The gravitational fields are evolved using the BSSNOK formulation with the same gauge conditions and parameters as in [3], while the GRMHD equations are solved using a high-resolution shock-capturing scheme based on the piecewise parabolic (PPM) reconstruction and the Harten-Lax-van Leer-Einfeldt (HLLE) approximate Riemann solver. For all the simulations discussed here we have used a grid structure with four refinement levels, an outer boundary located at $L \simeq 100\ M_\odot$ ($\simeq 150$ km) and finest resolution of $\Delta x = 0.375\ M_\odot$ ($\simeq 0.5$ km). With this setting the outer boundary is far enough to have all the dynamics happening inside the computational domain, while the used resolution is already able to capture the dynamics of the magnetic field (see [4]).

4. Results

For each unstable model (`U3`, `U11` and `U13`) we have performed a number of simulations adding an initial purely poloidal magnetic field around the typical value of the field strengths expected for magnetars (that is of the order of 10^{15} G) and, indeed, in the range from 1.0×10^{14} G to 1.0×10^{16} G. Much stronger initial magnetic fields are quite unlikely to be realistic, while weaker fields seems to have no effects at all on the dynamics of the bar-mode instability.

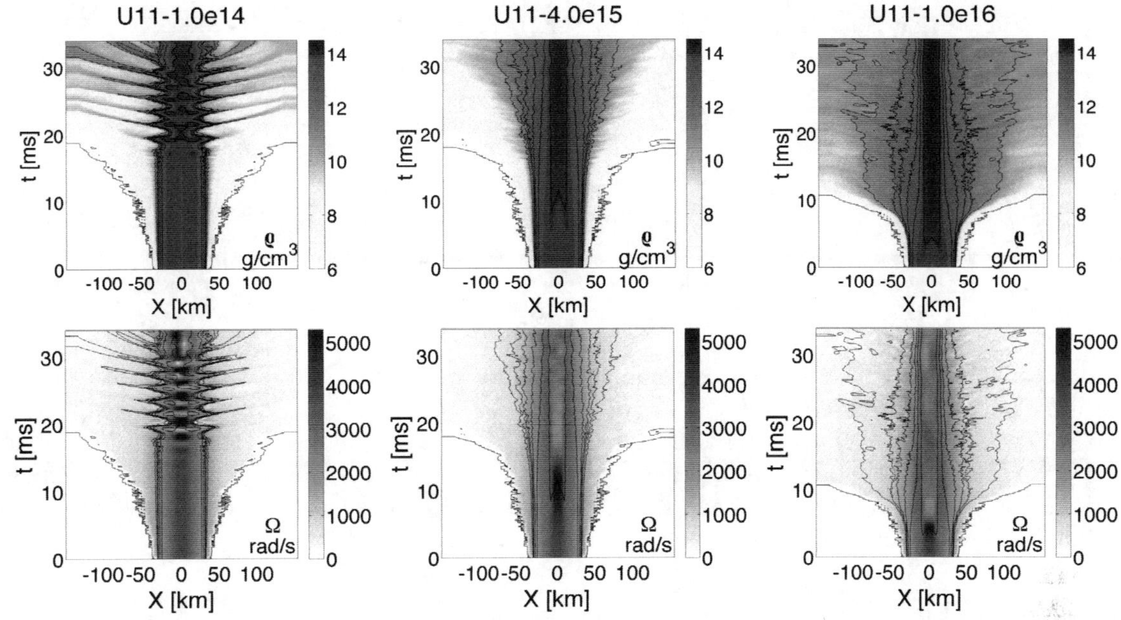

Figure 3. Evolution of the rest-mass density ρ (top row) and the angular velocity Ω (bottom row) along the x axis for models `U11-1.0e14` (left panels), `U11-4.0e15` (central panels) and `U11-1.0e16` (right panels). The panels show the color-coded quantities embedded in a space-time diagram with the coordinate time t on the vertical axis. Additionally, on top of all diagrams isodensity contours are shown for $\rho = 10^6$, 10^{11}, 10^{12}, 5×10^{12}, 10^{13}, 5×10^{13} and 10^{14} g/cm^3.

The dynamics of the bar-mode instability ($m = 2$ mode) is studied following the evolution of the distorsion parameter $\eta = (\eta_+^2 + \eta_\times^2)^{1/2}$, which is defined in terms of the matter quadrupole moment in the xy directions through the quantities $\eta_\times = 2\,I^{xy}/(I^{xx} + I^{yy})$ and $\eta_+ = (I^{xx} - I^{yy})/(I^{xx} + I^{yy})$. In the central panel of figure 1 we show how the dynamics of η is affected by different choices of the initial magnetic field strength: in unstable models, like model `U11-1.0e14`, this parameter shows an exponential growth (whose growth time, τ_{bar}, can be easly measured).

In particular, regarding the bar-mode instability, we draw the conclusion that there are three distinct behaviors corresponding to different seed magnetic fields.

Indeed, we observe no effects at all on the dynamics of the instability up to a certain magnetic field strength (namely, $B^z_{\rm max}|_{t,z=0} \lesssim 2.0 \times 10^{14}$ G for model `U3`, 4.0×10^{14} G for model `U11` and 8.0×10^{14} G for model `U13`), since both the growth time and the frequency of the m=2 matter mode do not vary within the accuracy they can be computed with (see the left panels of figure 2 and figure 3 and the right panel of figure 1). For higher field strengths we observe a gradual slowdown in the development of the instability, since the magnetic field lines are frozen into the fluid and the differential rotation has to drag them together with the matter implying that the instability growth time τ_{bar} is higher for higher initial values of $\beta_{\rm mag}$ (see the right panel of figure 1). The models that are still matter-unstable, even when a magnetic field is present, are represented with squares in the left panel of figure 1 showing the parameter space (β, $\beta_{\rm mag}$).

Going to much higher fields we reach a threshold value above which the instability is completely suppressed and the deformation no longer develops. This threshold on the magnetic field still depends on the initial model adopted, being higher than about 6.0×10^{14} for model `U3`, 2.0×10^{15} for model `U11` and 2.4×10^{15} for model `U13` (see figure 1).

Besides, for very strong seed magnetic fields, we also observe changes on the density and

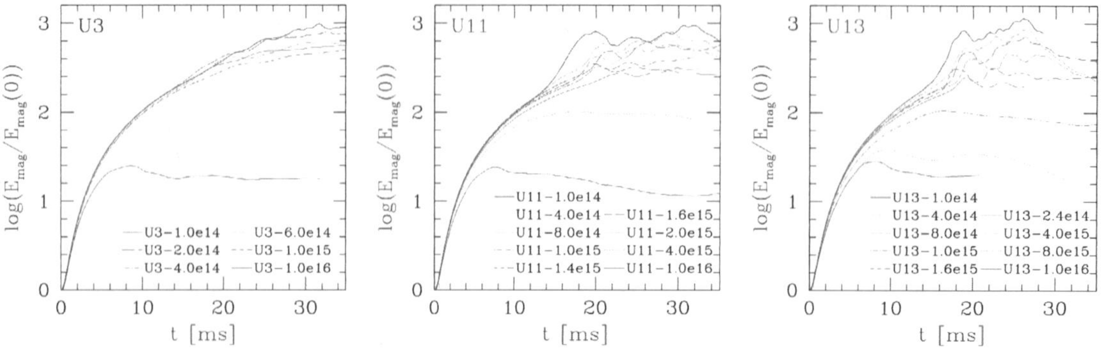

Figure 4. Time evolution of the total magnetic energy $E_{\rm mag}$ normalized to its initial value for the three models which are unstable in the non-magnetized case (U3, U11 and U13) corresponding to a wide range of seed magnetic fields from $B^z_{\rm max}|_{t,z=0} = 1.0 \times 10^{14}$ G to 1.0×10^{16} G. We use a black solid line to depict the less magnetized case, a blue solid line to indicate the most magnetized case and a red solid line for the last unstable model just before the suppression of the bar-mode instability due to the presence of the magnetic field.

angular velocity profiles of the stellar models, so in the end of the evolution we obtain a much more compact configuration that is nearly uniformly rotating and surrounded by a very low density envelope which is still differentially rotating (see the right panels in figure 2 for model U11). This is consistent with the expectation that magnetic braking is transferring angular momentum from the core to the outer layers. Models that are matter-stable are indicated by circles in figure 1, while the ones that also exhibit the above-mentioned expansion of the outer layers due to the presence of the magnetic field are represented with triangles.

Taking a closer look at the magnetic field evolution for all seed magnetic fields, we observe that magnetic field lines wind due to differential rotation, and hence a development and linear growth of a toroidal component take place, with a consequent amplification of the total magnetic energy of about two orders of magnitude or even more (see figure 4). Moreover, we observe a further exponential growth of both poloidal and toroidal components during the matter-unstable phase, whose nature still needs to be investigated.

After having studied how the magnetic field affects the onset and development of the bar-mode instability, we turned to investigate whether there are effects on the dynamics of models which are known to be stable in the non-magnetized case. Once again, this was achieved by superimposing a purely poloidal seed magnetic field allowing the magnetic field strength to take two values; $B^z_{\rm max}|_{t,z=0} = 1.0 \times 10^{15}$ G and $B^z_{\rm max}|_{t,z=0} = 1.0 \times 10^{16}$ G.

Regarding the distribution of matter we observe negligible effects when $B^z_{\rm max}|_{t,z=0} = 1.0 \times 10^{15}$ G. Only for model S1, which is very close to the threshold for the onset of the dynamical instability (see the left panel of figure 1), we observe some minor changes. More specifically, the density profile turns from a toroidal initial configuration to a nearly standard one with its maximum quite close to the z-axis. In stable models with $B^z_{\rm max}|_{t,z=0} = 1.0 \times 10^{16}$ G, instead, we observe the same behavior already described for highly magnetized models, namely the outer layers expand forming an envelope of much lower density around the initial stellar model.

5. Conclusions

We have analyzed how the presence of magnetic fields can affect the development of the bar-mode ($m = 2$) instability in relativistic differentially rotating stellar models with a seed poloidal magnetic field in the range $10^{14} - 10^{16}$ G. In order to do that we have performed 3D magneto-hydrodynamical simulations coupled to the Einstein equations.

Journal of Physics: Conference Series **470** (2013) 012008 doi:10.1088/1742-6596/470/1/012008

For all the models studied we have found, as expected, a sudden formation and linear growth of a toroidal component of the magnetic field (in a twisted configuration) that rapidly overcomes the original poloidal one and an amplification of the total magnetic energy to up to almost three orders of magnitude.

For relativistic stellar models that are bar-mode unstable, we have observed almost no effects on the bar-mode dynamics due to magnetic fields of the order of 10^{15} G or below. For stronger magnetic fields the growth time of the instability increases and the bar deformation appears to partially stall. Moreover, magnetic fields stronger than about 10^{15} G are able to completely suppress the instability and to change the density and angular velocity profiles of the stellar models, eventually leading to a final configuration where the star has almost uniform rotation and is surrounded by a very low density envelope.

he same evolution is realized also for relativistic stellar models that are stable against bar-mode deformations for seed magnetic fields of the order of 10^{16} G. Moreover, we note that the presence of magnetic fields does not alter the dynamics in stable models up to field strengths of the order of about 10^{15} G.

The overall conclusion is that only magnetic fields of the order of 10^{15} G or above seem to be able to suppress the purely hydrodynamical instability and so it is quite unlikely that in realistic astrophysical situations such a suppression due to the presence of magnetic fields might occur. Indeed, the general picture discussed in [1] applies also in the non Newtonian regime.

This research has been possible thanks to the HPC resources of the INFN "Theophys" cluster and to the PRACE allocation (6th-call) "3dMagRoI" on CINECA's Fermi supercomputer.

References

[1] Camarda K, Anninos P, Fragile P C and Font J A 2009 *Astrophys. J.* **707** 1610-22
[2] Rezzolla L, Giacomazzo B, Baiotti L, Granot J and Kouveliotou C 2011 *Astrophys. J. Lett.* **732** L6
[3] Baiotti L, De Pietri R, Manca G M and Rezzolla L 2007 *Phys. Rev.* D **75** 044023
[4] Franci L, De Pietri R, Dionysopoulou K and Rezzolla L 2013 *Preprint* gr-qc/1308.3989
[5] Giacomazzo B, Rezzolla L and Baiotti L 2011 *Phys. Rev.* D **83** 044014

Journal of Physics: Conference Series **470** (2013) 012009 doi:10.1088/1742-6596/470/1/012009

Modeling Galactic Extinction with dust and "real" PAHs

A Zonca[1], G Mulas[2], S Casu[2] and C Cecchi-Pestellini[3]

[1] Dipartimento di Fisica, Università di Cagliari, Strada Prov.le Monserrato-Sestu Km 0.700, I-09042 Monserrato (CA), Italy

[2] INAF-Osservatorio Astronomico di Cagliari, Strada n.54, Loc. Poggio dei Pini, I-09012 Capoterra, Italy

[3] INAF-Osservatorio Astronomico di Palermo, P.zza Parlamento 1, I-90134 Palermo, Italy

E-mail: azonca@oa-cagliari.inaf.it

Abstract. We elaborated an interstellar dust model assuming a distribution of core–mantle grains and a collection of single polyciclic aromatic hydrocarbons. Exploiting this model, we are able to reproduce a very large sample of galactic extinction profiles with very different flavours, proving that a polyciclic aromatic hydrocarbons population can reproduce extinction curve features in the ultraviolet range. Dust grains are composed by an hollow silicatic core and a carbonaceous mantle (this description is mutuated by the cycle of carbon in the interstellar medium); molecular population is represented by 54 molecules in four charged states and, despite the large number of free parameters (we have 9 parameters to represent grain distribution and 54×4 column densities to reproduce the molecular contribution to the extiction), we are able to determine some global properties for molecular ensemble and we found that these properties are indipendent by specific species which we use in our model.

1. Introduction

Interstellar dust grains are an important component of the interstellar medium (ISM). Deducing their nature is a very important step in the understanding of chemistry, dynamics and thermodynamics of the interstellar medium in the Milky Way and in external galaxies [1].

There are many lines of evidence of interstellar dust presence in the ISM, from which astronomers can deduce different dust properties. A strong indication for dust presence is the reddening of stellar light, caused by the absorption and scattering by dust grains. This reddening is wavelength–dependent, so an interstellar extinction curve can be derived by comparing the spectrum of an unreddened star with the one of a reddened star of the same spectral class. Other indicators of the nature of interstellar dust are: elemental depletions, polarization of stellar light, absorption and emission features [2].

The extinction curve is the observational tool most frequently used to infer dust properties. There is a basic similarity in the shapes of these curves: a general rise in normalised extinction from low values in the infrared to high values in the far ultraviolet, a near linear portion in the optical region, and a pronounced and broad "bump" near 217.5 nm. Hypothesis on the origin of this feature require Carbon, due to the presence, in the same spectral range, of the electronic transition $\pi^* \leftarrow \pi$ typical for aromatic carbon. By modelling interstellar extinction curves, we can deduce the grain size distribution. By depletion data we know that important components in interstellar dust are silicates [3] and carbonaceous materials, both in graphitic (sp^2) and in polymeric (sp^3) valencies [4].

During the last thirty years, different models have been elaborated, mostly based on the model by [5] in which dust grains are represented with a double population (graphite and silicate grains) following a power law size distribution (the Mathis-Rumpl-Nordsieck "MRN" distribution).

A model with significant differences from [5] model has been developed by researchers at Cagliari and Messina [6, 7]. In this model, astronomical silicates and solid carbon are present, but these components form a single population of dust grains in which carbon is assumed to be deposited on bare silicatic cores following description of [4] model. According to this model, dust grains are composed by bare silicatic cores, covered by a layer of solid carbon with a rate depending on gas density and temperature. This carbon, initially hydrogen-rich and sp^3 bonded, is then converted to sp^2 bonding at a rate (the so-called photodarkening rate) proportional to the intensity of the radiation field. The other component, which completes this model, is represented by Polycyclic Aromatic Hydrocarbons (PAHs), which can be modeled with a double Lorentzian or with "real" cross–sections for a sample of these molecules. Using the model in its simplified version (two Lorentzian profiles to represent PAHs contribution) [8] were able to fit a wide variety of peculiar galactic extinction curves.

In this paper, which is mainly based on [9], we use the model both in its simplified version and in including "real" PAHs cross-sections to fit the whole sample of extinction curves parametrized by [10]. We present the model in section 2 , a collection of results in section 3 and we draw our final remarks in section 4.

2. The model

Our interstellar dust model has two components

- classic component, represented by a distribution of silicatic core grains covered by a carbonaceous mantle;
- molecular component, represented by a population of Polyciclic Aromatic Hydrocarbons (PAHs).

We developed the model in two ways, differing only in the molecular representation (see section 2.2): in the simplified version we use a double Lorentzian approximation to reproduce the molecular contribution, while in the detailed version we use photo–absorption cross–sections of 54 PAHs in 4 charged states.

2.1. Classic component

This component is the same for both model versions and it is represented by dust grains with four concentric spherical layers: the core which is composed by a central void, to simulate the porosity, covered by a silicatic shell, and a double carbonaceous mantle with the inner layer composed by carbon sp^2 bonded and the outer layer composed by carbon sp^3 bonded. This particular composition takes into account the carbon galactic cycle in the interstellar medium according to the description by [4] and the cosmic abundances inferred by gas phase elements depletions.

We assumed that the mantle thickness must be the same over the whole grain distribution as the mantle accretion in the interstellar medium suggests [11]. Likewise, the vacuum fraction in volume is a variable fit parameter, but it is indipendent of grain size and it is the same for all grains. We adopted a power law distribution for dust grains size $(a + w)^{-q}$, in which a is the outer silicatic core radius and, in our model, it ranges from 5 to 1000 nm, w is the mantle thickness and it is free to vary from 0 to 3 nm, q is the exponent of the power law distribution, it is a free parameter and its value is around 3.45. This distribution allows for a gap in particle size dividing small grains by large ones, hence characterizing them by a lower and upper size limit. These two populations have no qualitative differences and the components are the same for both.

2.2. Molecular component

By the particular shape of PAHs cross–sections from UV to FUV range, we can infer that these molecules can contribute to the interstellar extinction in that spectral range. Futhermore these molecules are responsible for IR emission bands which are observable along different lines of sight and their contribution to the extinction is consistent with the observed emission in the IR bands.

In our model PAHs are represented in two ways: i) with a double Lorentzian, ii) with "real" cross–sections of a mixture of 54 PAHs in four charge states.

In the simplified version, the PAHs contribution to the extinction is approximated with a double Lorentzian profile following the description of [6]. The first Lorentzian mimics the $\pi^* \leftarrow \pi$ resonance in PAHs and reproduces the characteristic bump at 4.6 μm^{-1} of interstellar extinction curve; the second Lorentzian mimics the low energy tail of the $\sigma^* \leftarrow \sigma$ resonance, reproducing the far ultraviolet non linear rise in the interstellar extinction curve. An exponential cut–off on the low energy tail of the $\sigma^* \leftarrow \sigma$ resonance has been introduced to take into account the lower contribution of PAHs to extinction at low energies compared to the contribution of double Lorentzian profiles. We calibrated double Lorentzian on cross–sections of a mixture of 54 PAHs in four charge states: 0, $\pm$ 1, +2.

In the detailed version, we used a linear combination of computational photo–absorption cross–sections of a sample of 54 PAHs in the same charge states which we used to calibrate double Lorentzian. In this representation, a number of 216 free parameters (4 charge states for each PAH) have been considered.

3. Results

Using both the model in the simplified version and in the detailed version, we fitted the full set of [10] interstellar extinction curves. We used a modified version of C MPFIT implementation of Levemberg–Marquardt non linear least squares algorithm to perform the fits. This iterative procedure seeks to minimize numerical χ^2 starting from a first guess of free parameters, staying on parameters constraints. A possible problem with this procedure is that our fit could fall on the nearest minimum which could be reached in a saddle point. Therefore, to avoid a similar problem and to evaluate the covariance matrix, after performing a fit on original data, we perturb them with a Gaussian noise proportional to the errors and we performed a fit on these perturbed data. In some cases the perturbed fit was better than for unperturbed data, showing that the first fit falled on a local minimum or a saddle point.

Figure 1 shows two interstellar extinction curves representative for the whole set and their fit using the detailed model. The model is able to fit the remarkable variety of interstellar extinction curves: from bumpless curves to curves similar to the galactic average one. Therefore PAHs could be considered as a solution to reproduce ultraviolet extinction features.

Some of the classical parameters are constant across the whole sample of fitted curves. This is the case for the exponent of the power law distribution, which is almost always very close to 3.45, and for the minimum size for small grains (it is nearly always fixed on the limit value of 5 nm, then the fit, if unconstrained, would in most cases try to move towards smaller grains). The upper limit for large grains size shows little variations, suggesting that our fit is insensitive to the population of bigger grains. Carbonaceous mantles are mostly 1 nm thick and they are mostly aromatic. In some cases we obtain bare silicatic grains (only with simplified model): in these lines of sight we expected that very small grains would emit in the silicate bands when transiently heated by the absorption of energetic photons, but such emissions are not observed. Futhermore a small number of lines of sight has mantle tickness values of 3 nm (our upper limit).

Journal of Physics: Conference Series **470** (2013) 012009 doi:10.1088/1742-6596/470/1/012009

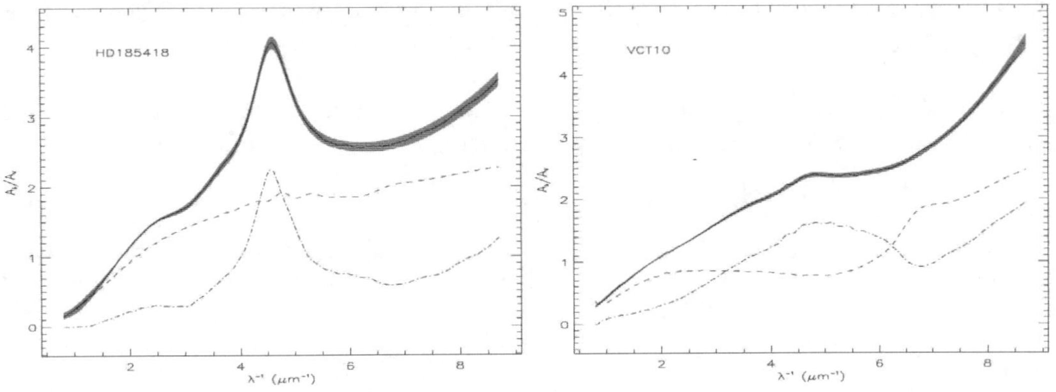

Figure 1. Fits to extinction curves for two lines of sight: HD185418 and VCT10. Solid lines: fit; dashed lines: classical dust contribution; dot-dashed lines: molecular contribution. The shaded area represents the observational errors given by [10].

It is likely that, if left free, these fits would have converged to somewhat larger values for mantle thickness.

Regarding the molecular component, we are able to determine from our fits some global properties for PAHs, for example column density and average charge for the PAHs mixture. On the other hand the PAHs detailed composition is largely unconstrained by the fit, so our results are not dependent by the specific species used for the fit. Although the two versions of the model have the same classical component and double Lorentzian, in the simplified version, are calibrated on the same PAHs mixture photo–absorption cross–sections that we used in the detailed version, comparing fit results of the two versions for the same line of sight, we found some differences both in PAHs abundances and in classical part. This is due by the different flexibility of the two versions of our model: i) the two Lorentzians have complete freedom to vary their relative intensity but their shape must be strictly Lorentzian, ii) PAHs can vary the particular shape of their extinction contribution but the intensity ratio between two resonances has limited variability. As a consequence, when we fit the same extinction curve, especially if it is a peculiar one, classic dust must compensates these differences on molecular contribution to the extinction (see figure 2), resulting in the observed difference in classical dust distributions between the two models.

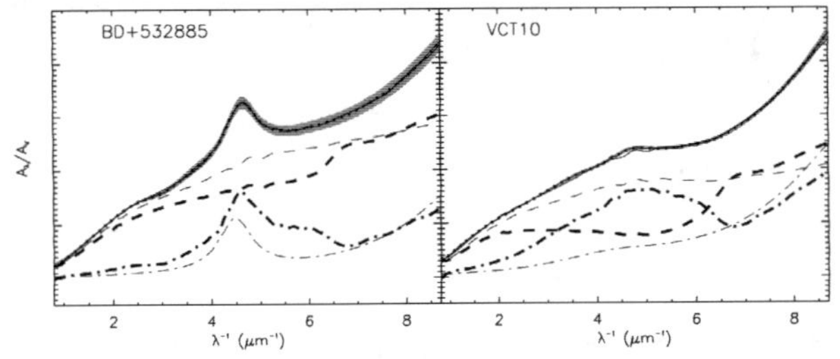

Figure 2. Fits to extinction curves for two lines of sight: BD+532885 and VCT10. The simplified fit is plotted with thin lines, the detailed one with thick lines. Solid lines: fit; dashed lines: classical dust contribution; dot-dashed lines: molecular contribution. The shaded area represents the observational errors given by [10].

We also searched a correlation between the average charge per carbon atom and the ratio of c_4 and c_3 parameters of parametrization introduced by [10]. These two parameters represent the FUV rise intensity and the bump intensity respectively. Indeed it is known that PAHs with an increasing positive ionization have a larger gap between the $\pi^* \leftarrow \pi$ and the $\sigma^* \leftarrow \sigma$ resonances

[7]. We found a positive correlation with some scatter, which confirms that PAH charge plays a role in determining the ratio between bump and FUV rise intensities but it does not solely determine it.

The amount of required Carbon locked in both solids and molecules is consistent with the interstellar budget for F and G stars, but too much Si is required by our model to fit interstellar extinction curves (see Fig. 14 of [9])

4. Conclusions

With our model we are able to fit the whole sample of 328 galactic interstellar extinction curves of [10], using both the simplified version and the detailed version. Usually interstellar dust models are applied only to the average Galactic interstellar extinction curves or to a few individual lines of sight.

We determine some global properties for PAHs mixture, as average charge per Carbon atom and column density, but individual properties for the molecules are unconstrained. With our model we also find that bumpless interstellar extinction curves are not necessarily lines of sight devoid of PAHs and we do not need particular exotic dust properties to fit them.

Our fitting algorithm tries to reach the best match with observations regardless of abundance constraints, so we obtained that our model requires too much atoms (especially Si) to fit interstellar extinction curves. Actually, to reduce the required abundances, we are perfoming new fits with an increasing χ^2 penalty term for abundances increasingly exceeding constraints.

The next step will be to extend the analysis with our model to external galaxies, starting from closest galaxies as the Small and Large Magellanic Clouds, to learn more about the cosmic history of the dust.

Acknowledgements

A. Zonca gratefully acknowledges Sardinia Regional Government for the financial support of his PhD scholarship (P.O.R. Sardegna F.S.E. Operational Programme of the Autonomous Region of Sardinia (Italy), European Social Fund 2007-2013- Axis IV Human Resources, Objective l.3, Line of Activity l.3.1.).

References

[1] Williams D A 2000 *Astronomy and Geophysics* **41** 030000
[2] Jones A 2009 *EAS Publications Series* **34** 107
[3] Li A and Draine B T 2001 *ApJ* **550** L213
[4] Jones A P, Duley W W and Williams D A 1990 *QJRAS* **31** 567
[5] Mathis J S, Rumpl W and Nordsieck K H 1977 *ApJ* **217** 425
[6] Iatì M A, Saija R, Borghese F, Denti P, Cecchi-Pestellini C and Williams D A 2008 *MNRAS* **384** 591
[7] Cecchi-Pestellini C, Malloci G, Mulas G, Joblin C and Williams D A 2008 *A&A* **486** L25
[8] Zonca A, Cecchi-Pestellini C, Mulas G and Malloci G 2011 *MNRAS* **410** 1932
[9] Mulas G, Zonca A, Casu S and Cecchi-Pestellini C 2013 *ApJS* **207** 7
[10] Fitzpatrick E L and Massa D 2007 *ApJ* **663** 320
[11] Whittet D C 2002 *Dust in the galactic environment* (Bristol: Institute of Physics Publishing)

Journal of Physics: Conference Series **470** (2013) 012010 doi:10.1088/1742-6596/470/1/012010

An accurate determination of the slope of the UV continuum in $z \sim 3$ star-forming galaxies

Stefano Pilo

INAF - *Osservatorio Astronomico di Roma, Via Frascati 33, 00040 Monteporzio (RM)*

E-mail: stefano.pilo@oa-roma.inaf.it

Abstract. Measuring dust extinction in high-redshift galaxies is fundamental to estimate the star formation rate density and constrain galaxy evolution models. I measure dust extinction in Lyman break galaxies at $z \sim 3$ where a broad peak in the star formation rate density is thought to occur. On the basis of image simulations I perform a new method for accurate extraction of colours from images with different point spread functions. Taking advantage of a new colour selection criterion I select a sample of 225 LBGs at redshift $< z > \sim 3.2$. The UV slope of the sampled galaxies is determined by a linear fit of the observed magnitudes through a proper set of filters coherently with the original definition of the slope β. By performing measurements of β through IZY filters I find $< \beta > \simeq -1.79$; moreover I find that the β value strongly depends on the set of filters I use for linear fitting. I find a good agreement comparing the three bands fit in our sample with CANDELS GOODS-South data set at the same redshift. UV slope value fits well in the trend for which dust extinction increase at decreasing redshift.

1. Introduction

Since dedicated Lyman Break Galaxy (LBG) surveys has been composed about 20 years ago (e.g. [1,2]), a method called "Lyman break technique" has been improved in order to detect star-forming, moderately absorbed galaxies. Thanks to technical developments that have allowed to reach fainter magnitudes, LBGs have been detected at ever higher redshifts in a wide range of wavelengths. Surveys of this kind are very useful to explore statistical properties of high redshift galaxies in order to measure some fundamental physical quantities like star formation rate density (SFRD) [3], luminosity function (LF) [4, 5] and mass function (MF) [6] to better understand the dynamics of their evolution through cosmic time. Measuring SFRD relies on a proper conversion between the rest-frame UV luminosity and the ongoing star formation rate (SFR). Evaluating the slope of the UV range of a galaxy spectrum is a powerful tool to constrain physical properties that influence this conversion. This work concerns the dust extinction as the most important factor affecting the shape of the rest-frame UV spectrum. Assuming that dust absorption is the strongest reddening factor, the slope of the UV spectrum can be directly related to dust extinction of a galaxy assuming that the spectrum of a star forming galaxy can be well approximated by a power law longward the Ly-α wavelength until 3200Å as discussed in previous works (e.g. [7, 8]). This power law can be written as $F_\lambda = \lambda^\beta$ where β is actually the slope I want to measure and it is equal to about -2.2 for a dust-free stellar population with solar metallicity and constant SFR. Even if many galaxy samples have been collected up to now, β measurement is still affected by large uncertainties (e.g. [3], [9–11]). Discrepancies among different results can be explained by ineffective selection criteria and by the method used to measure the UV slope itself, often performed by using a single colour. In this paper I will show a possible way to avoid these limitations including infrared bands both in selection criteria and in the slope measurement: this enhances the efficiency of the selection criterion and can provide a more "stable" fit of the observed magnitudes. Moreover, I will present an original tailored method to perform aperture photometry when dealing with images with different PSFs. The structure of the paper is as follows: in section 2, 3 and 4 I will present dataset and two original methods I used for photometry and selection criteria; in section 5 results will be shown with an overview on critical points in the linear fitting through different sets of filters; summary and conclusions (section 6) will follow at last.

2. Dataset

For this work photometry has been extracted using SExtractor software [12]. I exploited 9 different images in as many different filters acquired in the COSMOS field [13]. Images have been acquired by LBC camera on Large Binocular Telescope (U_{035}, G_{047}, R_{062} having central wavelength 355nm, 475nm, 622nm respectively), Suprime-Cam on Subaru Telescope (I_{076}, Z_{090} having central wavelength 764nm and 903nm respectively) and Vir-Cam on Vista Telescope (Y_{102}, J_{125}, H_{165}, K_{215} having central wavelength 1020nm, 1250nm, 1650nm, 2150nm respectively). The total area covered by these observations is $\sim$ 504 sq. arcmin. The measured limiting magnitudes for each image are the following: $U = 26.4$, $G = 26.8$, $R = 26.2$, $I = 26.0$, $Z = 24.8$, $Y = 24.8$, $J = 24.4$, $H = 24.2$, $K = 23.2$ at 10σ. A multiwavelength catalogue of the sources has been compiled using the R_{062} band as detection band.

3. A new method for accurate ground based photometry

Since the images have been acquired using different telescopes under different weather conditions, they have different PSFs. For this reason the shape of the same object in each image is smoothed in a different way with respect to the others. This implies that fixed aperture photometry is not the right choice because apertures with the same diameter don't recover the same flux fraction from sources in different images, that is in different bands. To avoid this problem I re-define a sort of *effective FWHM* ($FWHM_{eff}$) that depends, for each image, only on the flux that is measured from the sources. This flux fraction is measured on bright stars in the field and it is arbitrarily assumed to be $\frac{FLUX\_BEST}{FLUX\_APER} = 1.5$; this fixes the aperture recovering 66% of the total flux. So the diameter of the aperture that recovers this fraction of flux is the new measure of the PSF of the image, that is what I have previously called the $FWHM_{eff}$. Once I measured for each image its $FWHM_{eff}$, I make extended simulations using an high resolution image acquired by HST in GOODS field [13] (H_{160} image with $FWHM$=0.18arcsec): the idea that lies behind these simulations is that when smoothing a single input image to two images with different PSFs, their colour must be equal to zero (when we collect the same fraction of flux for each object) by definition, because the images have been generated by the same one. Making many tests comparing colours between images smoothed to different PSFs, I find, for every combination of images I have simulated, what are the right apertures to be applied to the measure images in order to have colours equal to zero and to maximize the signal to noise ratio. Tests lead to an empirical relation that allows to choose the right dimension of the aperture for the measure image in terms of its own $FWHM_{eff}$. The aperture for the detection image is fixed to be twice its $FWHM_{eff}$ corresponding to the aperture that has been proved to maximize the signal to noise ratio. So I apply the above mentioned relation, that is

$$Aper(meas) = [FWHM_{eff}(det) - FWHM_{eff}(meas)] + 2 \tag{1}$$

where det refers to the detection image and $meas$ to the measured one. Thus I compile a catalogue of sources: total magnitude is measured as

$$MAG_{tot}(U) = MAG_{tot}(R) + (MAG(U) - MAG(R))_{aper} \tag{2}$$

where colours are measured in a variable circular aperture according to equation 1. Recalling that the R band is the detection one, equation 2 is applied to all the other filters.

4. A new selection criterion

It has been proved that modifying usual colour-colour selection criteria for $z \sim 4$ galaxies including a redder band H instead of the usual one Z, provides a cleaner selection of LBGs in the selected redshift range [3]. To extend this approach to lower z, I make a comparison on the basis of stellar populations libraries between the usual *(U-G) vs (G-R)* selection criterion and a new one *(U-G) vs (G-Y)* for $z \sim 3$ galaxies. The two different criteria are plotted in figure 1.

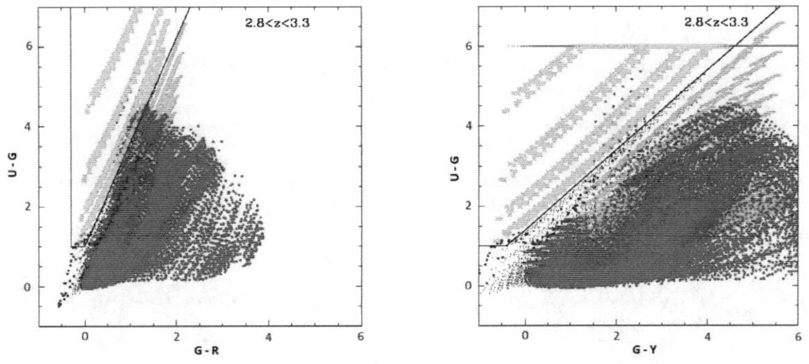

(a) UGR selection criterion. (b) UGY selection criterion.

Figure 1: Comparison between UGR criterion (panel a) and UGY criterion (panel b). Different colours stand for the evolution of different kinds of galaxies in the diagram: LBGs in the labeled redshift range (cyan), LBGs out of the desired range (black), passively evolving elliptic galaxies at lower redshift (red), dusty galaxies (green), stars (blue). The lines indicate the adopted selection window.

Selection cuts of the UGY criterion have been empirically defined by models taken from [14, 15]. They have been chosen as follows, by looking at the position in the UGY diagram of the sources in the redshift range of interest:

$$(U - G) > 1$$
$$(U - G) > (G - Y) + 1.24$$

The main drawback I want to avoid in this kind of selections is the contamination of the sample by lower redshift interlopers. This contamination arises because very reddened galaxies and passively evolving ones at lower redshifts can mimic the same spectral features of $z \sim 3$ LBGs that I actually want to select. The comparison between selection criteria shows that the UGY one provides a better and cleaner selection than the usual UGR, avoiding in a better way spurious sources from the selection window in the redshift range of interest. The chosen selection criterion provides a robust sample of 225 galaxies with the bulk of them having absolute magnitude between -22.6 and -20.8 and redshift between $z_{min} \sim 2.8$ and $z_{max} \sim 3.6$ providing a distribution with a proved photometric redshift $< z > \simeq 3.2$.

5. Estimating the UV slopes

Once the sample has been acquired by applying the UGY selection criterion to real data, β is measured by fitting a linear relation through a proper set of filters in order to sample the rest frame UV spectral range of the sources. Figure 2 shows the result obtained in measuring the UV slope of the sample through a 3-filters set (IZY) and the distribution of β itself. I obtain $< \beta >= -1.79 \pm 0.05$ where the error is calculated as the r.m.s. value of the distribution divided by the square root of the number of sources that form the selected sample. Assuming the Calzetti law [16], this value corresponds to a mean reddening $< E(B-V) > (z \sim 3) = 0.44(1/4.39)(2.31 < \beta > +4.85) \simeq 0.07$. Moreover, the plot shows a modest dependence of β on the absolute magnitude of the sources estimated at 1600Å rest-frame, in the sense that for fainter objects its value seems to increase. Assuming that the slope of the UV spectrum is largely dominated by the presence of dust in a galaxy, this result fits well in the literature trend in which dust extinction in high redshift galaxies increases when redshift itself decreases (e.g. [3, 11, 17, 18]). In order to test the results with a fainter magnitude sample I make a comparison with the CANDELS-GOODS field dataset. This consists of 296 galaxies in the same redshift range as the sample in the COSMOS field. As it can be seen in figure 3, the comparison with GOODS field sample shows a pretty good agreement with my β measurement: I obtain $< \beta >= -1.81 \pm 0.06$ for a linear fitting through I_{775}, Z_{850} and Y_{105} filters (with central wavelengths 771nm, 888nm and 1058nm respectively).

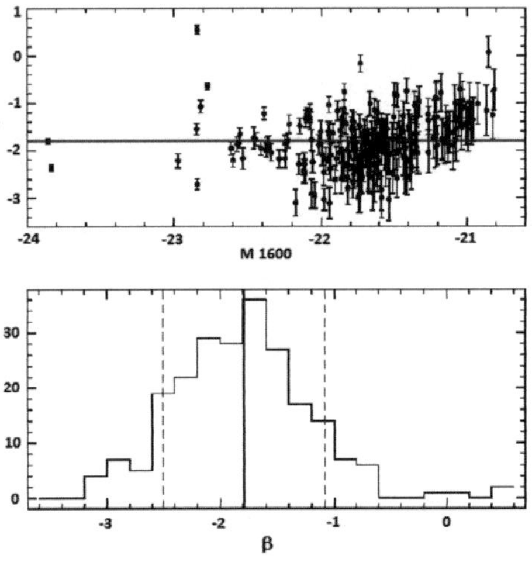

Figure 2: *Upper panel*: β vs rest-frame magnitude at 1600Å; the red line marks $< \beta >$. *Lower panel*: distribution of β where vertical central line is $< \beta >$ value and the dashed ones indicate $rms = 0.70$.

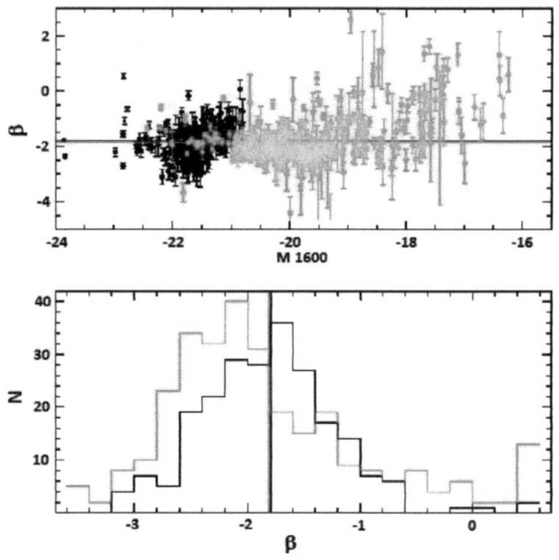

Figure 3: *Upper panel*: COSMOS sample β (black) and GOODS sample β (green) plotted together. *Lower panel*: the COSMOS and GOODS β distributions overplotted. Lines in both panels mark $< \beta >$ value for each dataset.

5.1. Considerations on filter choice

As already mentioned, an unbiased measurement of β is a matter of very precise and tricky photometry and it is tightly related to the range of the UV spectrum is actually chosen to be sampled, assuming it is well approximated by a power law. This comes out measuring β with a linear fit through the wider range of 5 filters ($RIZYJ$). This returns $< \beta >= -1.42 \pm 0.03$ that seems to indicate a too wide difference between values in literature. Moreover, looking at precedent works on $z \sim 4$ LBGs slope [3, 18], reporting $\beta(z \sim 4) \simeq -1.9$, I argue that this value of β would indicate a too strong evolution in the dust extinction between these two stages of cosmic time. Comparing these two different approaches in measuring the β slope of a sample, both advantages and drawbacks should be taken into account. A 5-filters measurement provides a more stable linear fit because it is constrained by 5 points; on the other hand, a 3-filter linear fit will be necessarily more noisy, increasing the error bars of each galaxy slope measurement.

As a counterpart instead, a 3-filter fit most probably samples the UV range of the spectrum in which the power-law approximation still stands in the whole range of redshift distribution of the sample. "Extreme" bands in the fit (such as R and J bands) can introduce biases. Indeed the R band could be (partially) affected by the presence of the Ly-α line, while the J band samples the UV-continuum at $\sim$ 3000Å rest-frame which is affected by the presence of old stellar populations.

6. Summary and conclusions

This paper presented a statistical study of high redshift LBGs. I have investigated their dust extinction by measuring the slope of the UV spectrum in a sample of 225 galaxies selected through a colour criterion in the redshift range $2.8 < z < 3.6$. The aim has been achieved by exploiting photometry extracted from a set of deep images in the optical-near infrared wavelengths in the COSMOS field. At a first stage a new method to measure colours from different PSF images has been applied in order to compile a catalogue of total magnitudes of the sources. Once the catalogue has been made, I exploited a new selection criterion based on UGY bands that have provided a robust sample of galaxies at $< z >\sim 3.2$. Finally the β slope has been measured by fitting a linear relation through a set of three filters (IZY) sampling the very central part of the rest-frame range $1500\text{Å} < \lambda < 3200\text{Å}$ within which the UV spectrum can be well approximated by a power law. I obtained $< \beta >= -1.79 \pm 0.05$ showing a modest dependence on the magnitude of the sources. This value is consistent with recent literature results and has been proved to be consistent with a fainter magnitude sample extracted from CANDELS-GOODS dataset; moreover it fits well in the well established trend for which dust extinction in galaxies increases when cosmic time runs from high redshifts to lower ones. Moreover this study has enlightened an interesting point in the β method itself of fitting a linear relation through a set of filters. Indeed I have found a strong dependence of the value of β on the set of filters used in the linear fitting procedure. Adding the R band at the very red part of the wavelength range and the J band at the bluer part I have found a value of $< \beta >= -1.42 \pm 0.03$: this probably relies on the fact that the two added extreme filters sample a part of the UV spectrum (partially) affected by some spectral features that influence the power law approximation and by the presence of old stellar populations. At the very end, although the value of β has been found to be consistent, an important caveat emerged, that is the necessity to sample, as accurately as possible, the same part of the UV spectrum when comparing the slopes at different redshifts.

References

[1] Steidel C and Hamilton D 1992 *A. J.* **104** 941
[2] Steidel C and Hamilton D 1993 *A. J.* **105** 2017
[3] Castellano M et al 2012 *A. & A.* **540** A39
[4] Steidel C C, Adelberger K, Giavalisco M, Dickinson M and Pettini M 1999 *Ap. J.* **519** 1
[5] Grazian A, Castellano M, Koekemoer A M et al 2011 *A. & A.* **532** A33
[6] Fontana A, Salimbeni S, Grazian A et al 2006 *A. & A.* **459** 745
[7] Calzetti D, Kinney A L and Storchi-Bergman T 1994 *Ap. J.* **429** 582
[8] Meurer G R, Heckman T M and Calzetti D 1999 *Ap. J.* **521** 64
[9] Bouwens et al. 2009 *Ap. J.* **705** 936
[10] Bouwens R J, Illingworth G D, Oesch P A et al 2010 *Ap. J.* **708** L69
[11] Finkelstein S L, Papovich C, Giavalisco M et al 2010 *Ap. J.* **719** 1250
[12] Bertin E, and Arnouts S 1996 *A. & As.* **117** 393
[13] Grogin N A, Kocevsky D D, Faber S M, et al *Ap. J.* **197** 35
[14] Bruzual A G, 2007 *IAU Symp. 241* ed A Vazdekis and R F Peletier p 125
[15] Bruzual G 2007 *From Stars to Galaxies: Building the Pieces to Build up the Universe* ed A Vallerani; R Tantalo and A Moretti *ASP Conf. Ser.* vol 374 p303
[16] Calzetti D, Armus L, Bohlin R C et al 2000 *Ap. J.* **533** 682
[17] Hathi N P et al 2013 *Ap. J.* **765** 88
[18] Bouwens R J, Illingworth G D, Oesch P A 2013 *Preprint* arXiv:1306.2950V1

Journal of Physics: Conference Series **470** (2013) 012011 doi:10.1088/1742-6596/470/1/012011

Energy consumption and entropy production in a stochastic formulation of BCM learning

L R de Oliveira, G Castellani and G Turchetti

Dip. di Fisica, Universitá di Bologna, Viale B. Pichat 6/2, 40126 Bologna IT

E-mail: lucianarenatadeoliveira@gmail.com

Abstract. Biochemical processes in living cells are open systems, therefore they exchange materials with their environment and they consume chemical energy. These processes are molecular-based and for that reason the role of fluctuations can not be ignored and the stochastic description is the most appropriate one. The chemical master equation describes in exact way the probabilistic dynamics of a given discrete set of states and helps us to understand and clarify the differences between closed and open systems. A closed system is related to a condition of detailed balance (DB), i.e. an equilibrium state. After a sufficiently long period, an open system will reach a non-equilibrium steady state (NESS) that is sustained by a flux of external energy. We demonstrate that two implementations of the BCM learning rule (BCM82) and (BCM92) are, respectively, always in DB, and never in DB. We define a one parameter parametrization of the BCM learning rule that interpolates between these two extremes. We compute thermodynamical quantities such as internal energy, free energy (both Helmholtz and Gibbs) and entropy. The entropy variation in the case of open systems (i.e. when DB does not hold) can be divided into internal entropy production and entropy exchanged with surroundings. We show how the entropy variation can be used to find the optimal value (corresponding to increased robustness and stability) for the parameter used in the BCM parametrization. Finally, we use the calculation of the work to drive the system from an initial state to the steady state as the parameter of the plasticity of the system.

1. Introduction

The BCM theory[1] was originally proposed to describe plasticity processes in visual cortex as observed by Hubel and Wiesel [1]. One of the main postulates of this theory is the existence of a critical threshold (the sliding threshold θ_M) and the BCM rule has been classically implemented in two ways: BCM82 [2] and BCM92 [3], depending on the definition of θ_M. In the BCM82, the neuron is assumed to be linear and θ_M is calculated as the power of the expectation value of the neuron postsynaptic activity c, that is $\theta_M = \langle c \rangle^2$ [2]. In order to have stable fixed points, the average used for θ_M in the BCM92 model is calculated with the square of the postsynaptic activity, $\theta_M = \langle c^2 \rangle$, as opposed to the original form, which squared the average of the output itself. If the expectation value of the output is zero, then the original form would have only trivial stable fixed points. This difference in the definition of θ_M leads to the possibility of deriving the rule from an energy function and of a statistical interpretation of these systems [3, 4].

[1] Named after E L Bienenstock, L N Cooper and P W Munro, the BCM rule is a physical theory of learning in the visual cortex developed in 1982.

Journal of Physics: Conference Series **470** (2013) 012011 doi:10.1088/1742-6596/470/1/012011

A "natural" way to cope with the BCM theory is the so-called Chemical Master Equation (CME) approach [5] that executes in a precise manner the probabilistic dynamics of a finite number of states, and recovers, in the thermodynamic limit ($N \rightarrow \infty$), the mean field approximation. The CME can be viewed as a Markov process describing the temporal evolution of the probability of a given discrete set of states [5]. The CME approach offers the possibility of computing the thermodynamic state functions of both closed and open systems (i.e. whether they satisfy or not the detailed balance condition). The detailed balance condition (DB) states that in equilibrium the sum per unit time of all transitions into any state n must be balanced by the sum of all transitions from n into other states n' [5]. An interesting observation is that the two implementations of the BCM rule can either satisfy or not the DB condition.

Let $\mathbf{m}$ be the synaptic weights and $\mathbf{d}$ the input signals received by the synapses, the BCM synaptic modification rule for a single neuron [2] has the form

$$\dot{m}_j = \phi(c, \theta_M) d_j \tag{1}$$

where m_j is the jth synaptic weight and d_j the jth input signal. The modification function $\phi(c, \theta_M)$ depends on the neuron activity level $c \propto (\ldots) \cdot \mathbf{d}$ (it is assumed a linear proportionality between the input $\mathbf{d}$ and the output c) and on a moving threshold θ_M, which is a nonlinear function of the cell activity history [6]. For low values of d ($d < \theta_M$), ϕ is negative; for $d > \theta_M$, ϕ is positive. The possibility of computing thermodynamical functions such as entropy, free energy and entropy production gives the opportunity to quantify the grade of "openness" of a system. Nevertheless the entropy variation can be related to the stability and robustness of the system.

2. Parametrization of the bidimensional case of the BCM rule

It turns out that the bidimensional version of BCM rule, with two orthogonal inputs is indicative of the general case of stochastic high-dimensional non-orthogonal inputs. Analysis that connects both has been given in [2, 3, 6]. The averaged version of the BCM learning rule, in the bidimensional case is:

$$\frac{d}{dt}\begin{pmatrix} m_1 \\ m_2 \end{pmatrix} = (PD)^T \begin{pmatrix} \phi_1 \\ \phi_2 \end{pmatrix} \tag{2}$$

where m_1 and m_2 are the synaptic weights, P is the diagonal matrix with the inputs probability of the inputs p_1 and p_2, D is the inputs matrix (a matrix whose rows are the input vectors $\vec{d_1}$ and $\vec{d_2}$), and the neuronal output in the linearity region is $c = \vec{m} \cdot \vec{d}$. The vector $\vec{\phi} = (\phi_1, \phi_2)$, is defined as $(\phi(\vec{m} \cdot \vec{d_1}, \theta), \phi(\vec{m} \cdot \vec{d_2}, \theta))$.

The BCM learning rule can be formulated in two ways: BCM82 [2] and BCM92 [3], based on two definitions of the moving threshold θ. It is possible to find a parametrization that interpolates with continuity between these two extremes by a suitable definition of θ_α

$$\theta_\alpha = \langle c^{1+\alpha} \rangle^{2-\alpha} \tag{3}$$

in this way we obtain for $\alpha = 0 \Rightarrow \theta_0 = \langle c \rangle^2$ and for $\alpha = 1 \Rightarrow \theta_1 = \langle c^2 \rangle$.

As was done in the work of Bienenstock et al. 1982 [2], we consider two special input vectors as $d_1 = (1, 0)$, $d_2 = (0, 1)$, with equal probability of appearing $p_1 = p_2 = 1/2$ and the definition of θ as in equation 3, in this way the the BCM rule becomes[2]:

$$\begin{aligned} \dot{m}_1 &= \frac{1}{2}\phi(m_1, \theta) = \frac{m_1}{2}(m_1 - \theta) \\ \dot{m}_2 &= \frac{1}{2}\phi(m_2, \theta) = \frac{m_2}{2}(m_2 - \theta) \end{aligned} \tag{4}$$

[2] A detailed description is given by Cooper et al. 2004 [6].

3. BCM rule and CME

Both systems (BCM82 and BCM92) can be studied by the CME because the number of synapses can be small, and the role of fluctuations can not be neglected. On the other hand, if these numbers increase, the CME approaches the deterministic equations (mean field limit). Another motivation for the CME approach is that we can state conditions for the validity of the detailed balance, and if we can compute the stationary distribution, we can also compute all the relevant thermodynamic quantities as free energy and entropy. The CME formalism allows us to deal with "discrete synapses", that are synapses with non-continuous values, as pointed out by several authors [2, 3, 4, 6, 7]. The CME for the system defined in equation 4 is:

$$\begin{aligned} \dot{p}_{m_1,m_2} &= (\mathbb{E}_{m_1}-1)r^{(m_1)}_{m_1,m_2}p_{m_1,m_2}+(\mathbb{E}^{-1}_{m_1}-1)g^{(m_1)}_{m_1,m_2}p_{m_1,m_2} \\ &+ (\mathbb{E}_{m_2}-1)r^{(m_2)}_{m_1,m_2}p_{m_1,m_2}+(\mathbb{E}^{-1}_{m_2}-1)g^{(m_2)}_{m_1,m_2}p_{m_1,m_2}. \end{aligned} \quad (5)$$

This CME is derived under the condition of a one-step Poisson process [5], $\mathbb{E}$ and $\mathbb{E}^{-1}$ are the forward and backward step operators, whose effect on an arbitrary function $f(n)$ is defined as $\mathbb{E}_n f(n) = f(n+1), \mathbb{E}^{-1}_n f(n) = f(n-1)$. The generation and recombination terms respectively are $g^{(m_i)}_{m_1,m_2} = \frac{m_i^2}{2N^2}$, $r^{(m_i)}_{m_1,m_2} = \frac{m_i\theta_\alpha}{2N^2}$, $i = 1, 2$, and N is the maximum value of the synaptic weight (proportional to the maximum number of molecules). To verify if the DB condition holds, we define a quantity that we call "commutator" $\mathcal{C}_\alpha(m_1, m_2)$, because it is the difference between the two possible paths (i.e. by joining the bottom left vertex with the upper right vertex) in an unitary square. The validity of this definition relies on the structure of the CME, which does not contain diagonal terms (i.e. there are no terms with simultaneous variations of m_1 and m_2)

$$\mathcal{C}_\alpha(m_1,m_2) = \frac{g^{(m_2)}_{m_1-1,m_2-1}\cdot g^{(m_1)}_{m_1-1,m_2}}{r^{(m_2)}_{m_1-1,m_2}\cdot r^{(m_1)}_{m_1,m_2}} - \frac{g^{(m_1)}_{m_1-1,m_2-1}\cdot g^{(m_2)}_{m_1,m_2-1}}{r^{(m_1)}_{m_1,m_2-1}\cdot r^{(m_2)}_{m_1,m_2}}. \quad (6)$$

If $\mathcal{C}_\alpha(m_1, m_2) = 0$, the DB condition always holds, whereas if $\mathcal{C}_\alpha(m_1, m_2) \neq 0$ the DB does not hold.

4. Thermodynamics quantities from CME:

In order to make a non equilibrium thermodynamics approach in terms of the chemical master equation, we have to know some concepts from the statistical mechanics of equilibrium. For the canonical ensemble one can define the Gibbs entropy S, the total energy U and the Helmholtz free energy F

$$S = -k_B\sum_{i=1}^N p_i \ln p_i, \qquad U = \sum_{i=1}^N p_i u_i \qquad \text{and} \qquad F = U - TS \quad (7)$$

In this paper we will assume that k_B and the temperature T are equal to 1.

Considering that the Gibbs entropy (equation 7) can be written as function of time and taking its time derivate[3]

$$\frac{dS(t)}{dt} = -\sum_{i=1}^N \frac{dp_i}{dt}\ln p_i - \frac{d}{dt}\sum_{i=1}^N p_i = -\sum_{i=1}^N \frac{dp_i}{dt}\ln p_i \quad (8)$$

we take into account the generic master equation $\frac{dp_i(t)}{dt} = \sum_{j=1}^N (p_j(t)q_{ij} - p_i(t)q_{ji})$, where q_{ij} are the transition rates and $p_i(t)$ is the probability that the system is in the i_{th} state. In equation 8

[3] The term $\frac{d}{dt}\sum_{i=1}^N p_i$ does not contribute since $\sum_{i=1}^N p_i = 1$.

the term $\frac{dp_i}{dt}$ can be replaced by the master equation and then we can write the time variation of entropy as [8, 9]

$$\frac{dS(t)}{dt} = -\frac{1}{2}\sum_{i,j=1}^{N}(p_j q_{ij} - p_i q_{ji}) \ln \frac{p_i q_{ji}}{p_j q_{ij}} + \frac{1}{2}\sum_{i,j=1}^{N}(p_i q_{ij} - p_j q_{ji}) \ln \frac{q_{ji}}{q_{ij}} \tag{9}$$

This derivation is in agreement with the classical definition of the entropy variation and may be written as the sum of two terms [10]:

$$\frac{dS(t)}{dt} = h_d - e_p \tag{10}$$

Where e_p is the entropy production rate, i.e. the entropy produced inside the system due to spontaneous process and h_d is the heat dissipation rate, i.e. the entropy supplied to the system by its surroundings. According to the second law of thermodynamics e_p must be zero for reversible (or equilibrium) transformations and positive for irreversible transformations of the system. The entropy supplied, h_d, may be positive, zero or negative depending on the interaction of the system with its surroundings [10].

We can still relate h_d and e_p to the thermodynamic variables at equilibrium. If we consider the definition of the Helmholtz free energy (equation 7), the entropy variation (equation 10) can be expressed as $\frac{dS(t)}{dt} = \frac{dU(t)}{dt} - \frac{dF(t)}{dt}$. Therefore, we identify

$$h_d = \frac{dU(t)}{dt} \quad \text{and} \quad e_p = \frac{dF(t)}{dt} \tag{11}$$

In this way we have the mathematical formulation for the thermodynamic variables in terms of the master equation. When the system presents detailed balance condition, for $t \to \infty$ it reaches an equilibrium state with $e_p = h_d = \frac{dS(t)}{dt} = 0$. In contrast to systems in equilibrium, systems in NESS present fluxes of physical quantities, such as particles or energy. Thus, DB is violated and energy is pumped into the system to sustain the NESS [9, 11, 12]: e_p and h_d are not null, but equal. This is necessary to ensure that S is finite asymptotically and $\frac{dS(t)}{dt} = 0$.

Using the result in equation 11, we can calculate the work to drive the system from an initial state to the stationary (or equilibrium) state as

$$\begin{aligned} \frac{dW_{h_d}}{dt} &= \frac{dU(t')}{dt'} - \frac{dU(\infty)}{dt'} \\ W_{h_d} &= \lim_{t\to\infty}\int_0^t (h_d(t') - h_d(\infty))dt' = \lim_{t\to\infty}\left[\int_0^t h_d(t')dt' - th_d(\infty)\right] \end{aligned} \tag{12}$$

and

$$\begin{aligned} \frac{dW_{e_p}}{dt} &= \frac{dF(t')}{dt'} - \frac{dF(\infty)}{dt'} \\ W_{e_p} &= \lim_{t\to\infty}\int_0^t (e_p(t') - e_p(\infty))dt' = \lim_{t\to\infty}\left[\int_0^t e_p(t')dt' - te_p(\infty)\right] \end{aligned} \tag{13}$$

The total work (work of entropy) is written as

$$W_S = W_{h_d} - W_{e_p} \tag{14}$$

Hence, we use the calculation of the work as the parameter of the plasticity of the system (where each value of e_p and h_d represents a probability configuration for the system).

Journal of Physics: Conference Series **470** (2013) 012011 doi:10.1088/1742-6596/470/1/012011

5. Results

In this section we are going to compare the behavior of the different formulations of the BCM rule, through the calculation of the thermodynamic variables. More specifically, we are interested in determining which model is more plastic, studying the differences between DB and NESS. We have to determine if systems with $\alpha \neq 0$ and $\mathcal{C}_\alpha(m_1, m_2) \neq 0$ are in a NESS condition. We can do this by the calculation of h_d, e_p and $\frac{dS}{dt}$, because as we enunciated in section 4 NESS is characterized by a sustained energy input and $h_d = e_p \neq 0$ in the stationary state. We simulate the time evolution of h_d and e_p for $0 \leq \alpha \leq 1$ and we verify that for $\alpha = 1$ (BCM92) the system presents $h_d = e_p = 0.0013465$ and for $\alpha = 0$ (BCM82) $h_d = e_p = 0$. We can infer that the BCM82 reaches an equilibrium state and the BCM92 reaches a NESS.

We will compute the work done by the system to reach the equilibrium configuration. To do this we compute W_{h_d}, W_{e_p} and W_S according to equations 12, 13 and 14 by numerical integration of the CME over a time span sufficiently long to reach the stationary distribution.

A first comparison between the two BCM models reveals that the work done to reach the stationary distribution is lower in the case of DB violation (see figure 1).

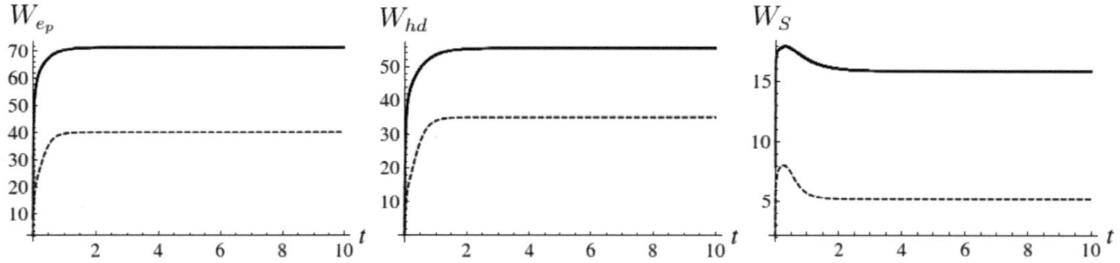

Figure 1. Change of W_{h_d}, W_{e_p}, W_S for BCM82 (black line) and BCM92 (dashed line). The simulations are performed for $N = 31$ and the initial condition $p_{m_1,m_2} = \delta_{m_1,31}\delta_{m_2,31}$

We confirm this trend by plotting the value of entropy work in the stationary state (W_S^s) as a function of α (see figure 2).

We note that the value of these quantities is dependent on the choice of the initial conditions. This dependence on the initial conditions is easily explainable for the W_S from the definition

$$W_S = \int_0^\infty \frac{dS}{dt'} dt' = \int_0^\infty (e_p(t') - h_d(t'))dt' = W_{e_p} - W_{h_d} = S(\infty) - S(0) : \qquad (15)$$

the entropy variation depends on the initial value. If for example we choose the initial conditions as: $p_{m_1,m_2} = \delta_{m_1,\bar{m}_1}\delta_{m_2,\bar{m}_2}$, the initial entropy is zero, hence $W_{e_p} - W_{h_d}$ is simply $S(\infty)$.

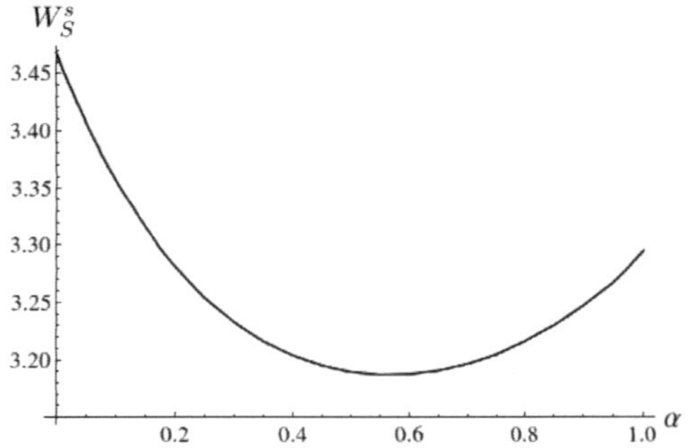

Figure 2. Change of W_S^s when α is varied from 0 to 1. The simulations are performed for $N = 31$ and the initial condition $p_{m_1,m_2} = \delta_{m_1,15}\delta_{m_2,15}$.

In figure 2 we show W_S^s as a function of the parameter α. It is possible to see that the entropy variation shows a minimum for $\alpha \approx 0.6$.

6. Conclusions

Our results show that when the system is not in the detailed balance condition, the work necessary to reach the stable state is less than that requested when the detailed balance holds. This means that the system requires less energy to memorize a pattern when the detailed balance is not satisfied. Hence the system is more plastic: a part of the energy that is requested to maintain the NESS is recovered when the system learns and develops selectivity to input pattern. We believe that this can be an hallmark of biological systems and that this can explain why these systems spend a large part of their metabolic energy to maintain NESS states; this energy is recovered during crucial developmental steps such as differentiation and learning.

References

[1] Hubel D and Wiesel T Receptive fields, binocular interaction and functional architecture in the cat's visual cortex. *Journal of Physiology*, **160**:106-154, 1962.

[2] Bienenstock E L, Cooper L N and Munro P W Theory for the development of neuron selectivity: orientation specificity and binocular interaction in visual cortex. *Journal of Neuroscience*, **2**:32-48, 1982.

[3] Intrator N and Cooper L N Objective function formulation of the BCM theory of visual cortical plasticity: Statistical connections, stability conditions. *Neural Networks*, **5**(1):3–17, 1992.

[4] Bazzani A, Remondini D, Intrator N and Castellani G C The effect of noise on a class of energy-based learning rules. *Neural Computation*, **15**(7):1621, 2003.

[5] van Kampen N G *Stochastic processes in physics and chemistry.* North Holland, 2007.

[6] Cooper L N, Intrator N, Blais B S and Shouval H Z *Theory of cortical plasticity.* World Scientific New Jersey, 2004.

[7] Castellani G C, Quinlan E M, Cooper L N and Shouval H Z A biophysical model of bidirectional synaptic plasticity: dependence on AMPA and NMDA receptors. *Proceedings of the National Academy of Sciences*, **98**(22):12772-7, 2001.

[8] Schnakenberg J Network theory of behavior of master equation systems. *Reviews of Modern Physics*, **48**:571–585, 1976.

[9] Ge H, Qian H Physical origins of entropy production, free energy dissipation, and their mathematical representations. *Phy. Rev. E*, **81**:051133, 2010.

[10] de Groot S R, Mazur P *Non-Equilibrium Thermodynamics.* North-Holland, 1962.

[11] Xu L, Shi H, Feng H and Wang J The energy pump and the origin of the non-equilibrium flux of the dynamical systems and the networks. *Journal of Chemical Physics*, **136** (16):165102, 2012.

[12] Qian H Open-system nonequilibrium steady state: statistical thermodynamics, fluctuations, and chemical oscillations. *J. Phys. Chem. B*, **110** (31):15063–74, 2006.

Journal of Physics: Conference Series **470** (2013) 012012 doi:10.1088/1742-6596/470/1/012012

Review of 2H-tetraphenylporphyrins metalation in ultra-high vacuum on metal surfaces

M Panighel[1,2], G Di Santo[1], M Caputo[1], C Lal[3,4], B Taleatu[3,5] and A Goldoni[1]

[1] ST-INSTM, Elettra-Sincrotrone Trieste S.C.p.A., SS 14 km 163.5, 34149 Trieste, Italy
[2] Università degli Studi di Trieste, P.le Europa 1, 34127 Trieste, Italy
[3] International Centre for Theoretical Physics, Strada Costiera 11, 34151 Trieste, Italy
[4] Centre for Non-Conventional Energy Resources, University of Rajasthan, Jaipur-04, India
[5] Department of Physics, Obafemi Awolowo University, Ile-Ife, Nigeria

E-mail: goldonia@elettra.eu

Abstract. The formation and conformational adaptation of self-assembled monolayer of 2H-tetraphenylporphyrins (2H-TPPs) on metal surfaces, as well as their metalation processes in ultra-high vacuum (UHV), are reviewed. By means of XPS, NEXAFS and STM measurements we demonstrate that, after the annealing at 550 K, a temperature-induced chemical modification of 2H-TPP monolayer on Ag(111) occurs, resulting in the rotation of the phenyl rings parallel to the substrate plane. Moreover, independently of the conformation, we report three different methods to metalate 2H-TPP monolayers in UHV. Experimental evidence indicates that the presence of a metal atom in the TPP macrocycle influences both the conformation of the molecule and its adsorption distance.

1. Introduction

The self-assembly of molecular building blocks on crystalline surfaces allows the creation of highly ordered nano-architectures, which cannot be achieved by conventional methods. In this context, organic-inorganic interfaces, due to their flexibility, are very promising for the realization of new compounds with controllable and tunable properties [1].

Porphyrins and similar macrocycle molecular systems are extremely suitable for this task, because they combine the ability to self-assemble with the presence of active sites of interaction for metals [2–4]. 2H-tetraphenylporphyrins (2H-TPPs), for example, can be deposited in ultra-high vacuum (UHV) in well-ordered monolayers on different substrates and, in the last years, it has been demonstrated that metal ions can be incorporated and stabilized inside the macrocycle with the substrate-mediated formation of metallo-porphyrins [5–31]. Actually, there is a large number of studies on this subject, in particular taking into account the role of the substrate metal atoms in structural and electronic properties of the formed adlayer and metallo-adlayer. For this purpose, synchrotron radiation techniques, such as X-ray photoemission spectroscopy (XPS) and near-edge X-ray absorption fine structure (NEXAFS) – due to their extreme chemical and electronic selectivity – combined with scanning tunnelling microscopy (STM), allow a deep understanding of the chemical and conformational properties of the adsorbed molecules and of their metalation on crystalline metal substrates.

Journal of Physics: Conference Series **470** (2013) 012012 doi:10.1088/1742-6596/470/1/012012

Here we review the formation of a monolayer of 2H-TPPs on several metallic substrates, the adsorption geometry and the conformation of the molecule and the possible ways to obtain metallo-porphyrins in UHV starting from the adsorbed free-base species.

2. Thermal-induced conformational adaptation

As shown in many experiments on 2H-TPP multilayer [8, 13, 25], as well as in gas-phase measurements [32], the presence of an intact macrocycle can be determined by photoemission spectroscopy, through which the existence of two atomic species of nitrogen (N) can be detected (iminic and pyrrolic). The pyrrolic N is bonded to an hydrogen atom, while the iminic is not, resulting in a binding energy (BE) difference in the N 1s core level spectrum of about 2 eV.

Actually, 2H-TPP molecules have been deposited on several metallic substrates, but Au [4, 19, 26] and Ag [5–7, 10, 18, 20, 33] are those on which it is possible to obtain a single layer without any modification of the macrocycle. In both cases, when deposited at room temperature (RT) or when the monolayer is obtained after the sublimation of a multilayer in the range $525 \div 580$ K, the XPS analysis shows the presence of two unaffected N 1s peaks, indicating that in the monolayer the molecule macrocycle is still protonated.

However, a remarkable effect on the phenyls adsorption geometry was observed on the monolayer (or sub-monolayer) of 2H-TPPs, after annealing above 525 K. On the Ag(111) and Au(111) surfaces, the macrocycle of 2H-TPP deposited at RT has been shown to be adsorbed parallel to the substrate surface, adopting a saddle shape conformation [29], whereas its phenyl legs exhibit a tilt angle of about $45° \div 55°$ with respect to the surface plane [29, 33]. Differently, after annealing above 525 K, Di Santo et al. [33] predicted and observed a temperature-induced conformational adaptation of the porphyrin molecules on the surface, resulting in the rotation of the phenyl rings parallel to the substrate plane.

As an example, figure 1 shows the NEXAFS spectra for two polarized beams at the N 1s (left panel) and C 1s (right panel) thresholds of the 2H-TPP monolayer adsorbed on Au(111) at room temperature (top) and after the annealing at 550 K (bottom). Regardless of the temperature, the N 1s spectra show enhanced or suppressed $\pi*$ features respectively for light polarized perpendicular or parallel to the surface. This suggests that the macrocycle, in which the N atoms are located, is adsorbed parallel to the substrate surface.

The C 1s NEXAFS spectrum in 2H-TPP can be seen as the superposition of the macrocycle and the benzene (phenyl) spectra [34]: in particular, the first $\pi*$ peak at 284.1 eV belongs to the macrocycle, whereas the large peak at 285.2 eV is mainly related to the phenyl $\pi*$ absorption spectrum.

For the monolayer deposited at RT (figure 1c) the macrocycle peaks follow the same behaviour as the N 1s NEXAFS spectra, again indicating that it is adsorbed parallel to the substrate surface. The large phenyl peak shows instead a very small dichroism suggesting that the phenyl rings have an average angle very close to 50°.

For the monolayer after the annealing at 550 K (figure 1d), the situation is completely different: all of the $\pi*$ peaks are suppressed when $\theta = 0°$, indicating that in this case the phenyl rings are parallel to the surface. This situation is actually quite strange because the rotation of the phenyl rings is inhibited by the steric interaction between the hydrogen atoms in the molecule.

A possible reaction and modification of the porphyrins able to explain this configuration is a dehydrogenation of eight carbon atoms (four in the phenyl rings and four in the macrocycle) with the formation of four new aryl-aryl carbon bonds. In this way, with the rotation of the benzene moieties, the molecule becomes flat and, therefore, reduces the distance from the substrate, making new and stronger bonds between their molecular π orbitals (both from phenyls and macrocycle) and the bands of the substrate. The details of this configuration can be described within the density functional theory (DFT) formalism [33].

The study of a sub-monolayer of CoTPP evaporated on Ag(111) demonstrated that the

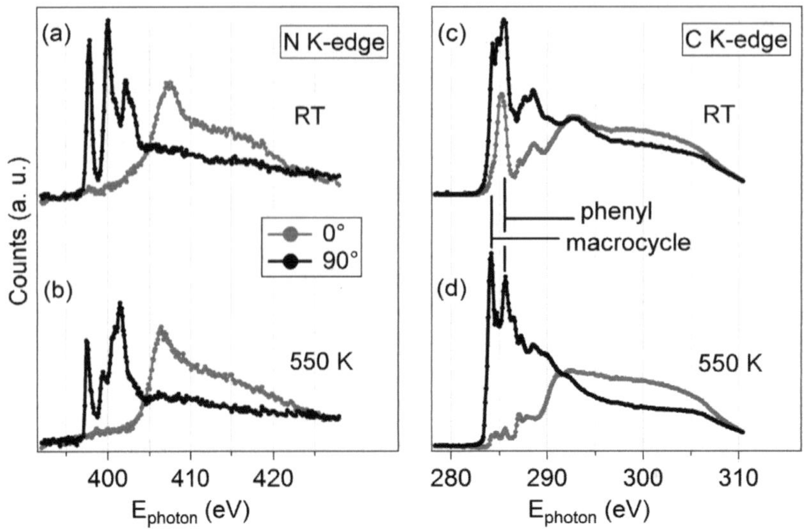

Figure 1. NEXAFS spectra for the angle of incidence $\theta = 0°$ (red) and $\theta = 90°$ (black) at the N 1s (left panel) and C 1s (right panel) absorption threshold for 2H-TPP monolayer deposited on Ag(111) at room temperature (top) and after annealing at 550 K (bottom).

molecule increases the number of bonds and reaches a more stable energetic configuration via this mechanism. The Co at the centre of the macrocycle, like other 3d metals (Fe, Ni, Mn, Cu), has a strong interaction with the substrate via the perpendicular d_{z^2} orbital, inducing the molecule to form bonds with the π orbitals. Indeed, after annealing up to 600 K, the CoTPP molecule does not change the orientation of the phenyl rings, that remain rotated by about 40° with respect to the macrocycle, which instead is parallel to the substrate surface [24]. A strong interaction with the substrate prevents this reconfiguration of the molecule; otherwise, when the interaction is weaker, like in the case of Zn, the rotation of the phenyl rings of ZnTPP is allowed as for the 2H-TPP molecule [35].

The hypothesis of 2H-TPP modification is supported by several recent analyses [12, 18], among which the most complete was by Papageorgiou et al. [18], who observed all the predicted dehydrogenation configurations through STM [33].

3. UHV in-situ metalation of the macrocycle

In-situ metalation of porphyrin molecules in UHV is of great interest for the characterization of pure species in a controlled environment. Generally the inclusion of metal atoms in porphyrin molecules is obtained in solution via wet chemistry, and typically most of the metallo-porphyrin molecules need a ligand coordination to the metal atom in order to reduce its reactivity and prevent oxidation. The possibility to produce in-situ stable pure metallo-porphyrins without any ligand and to characterize them is an intriguing challenge of fundamental interest. Furthermore, the capability of this type of molecules to bind various metal ions opens the discussion on how the presence of this central metal atom modifies the interaction with the substrate as well as on the modeling of the metalation process.

Several metals can be efficiently coordinated to 2H-TPP adsorbed on Ag(111) and Au(111) by in-situ metal evaporation: Fe [6, 25, 28], Co [8, 24], Zn [7], Ni [26], K [36] Ce [9], Mn and Rh [20] can be easily inserted in the macrocycle to form metallo-porphyrins. The metals can be coordinated either by direct evaporation on the 2H-TPP monolayer or multilayer, or by deposition of metal adatoms on the substrate before the molecules evaporation.

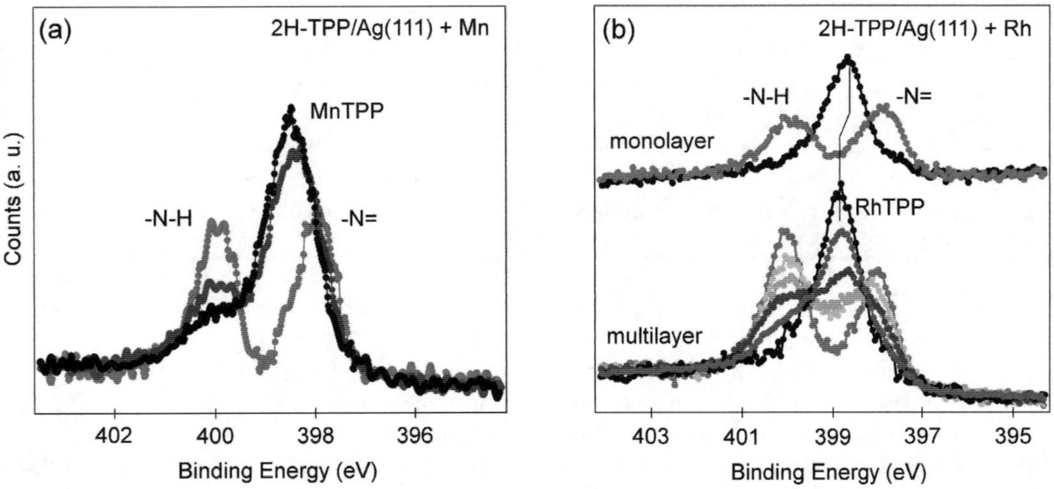

Figure 2. N 1s XPS spectra of 2H-TPP monolayer on Ag(111) during Mn (a) and Rh (b) evaporation (red spectra refer to the pristine 2H-TPP and black spectra to the metalated molecule). For the Rh case the metalation of the multilayer during consecutive evaporation steps is also shown.

Photoemission spectroscopy is extremely sensitive to the redox-reaction occurring at the macrocycle which leads to the metalation process. The evidence of the metal coordination comes from the changes of the N 1s XPS spectrum of the porphyrin when a metal ion is bound at the centre of the macrocycle. The spectrum of the 2H-TPP has two easily resolved components because of the two N species: the one at higher binding energy is assigned to the two pyrrolic N atoms (-NH-), while the lower BE peak corresponds to the two iminic ones (-N=); in the metallo-porphyrin, instead, the four N atoms are equivalent, and just one peak is expected.

Two examples, relative to the Rh and Mn metalation, are shown in figure 2. It is worth noting that the geometrical adaptation of the tetraphenylporphyrin in the monolayer has no implications on the possible hypothesis for the metalation mechanism. The metalation of other macrocycle molecules, with different geometries, is indeed happening in a similar way [5, 24–26, 30]. Therefore, a simple ad-atom hopping instead of a substrate surface-mediated diffusion to the reaction sites could explain the metalation process.

Recently it was shown that on some reactive substrates with a small ionic radius, like the first row 3*d* transition metals Fe and Ni [20], Cu [12, 14, 17, 22, 23] or Al [36] substrates, the 2H-TPPs in a monolayer metalate simply by picking-up substrate atoms or ad-atoms. In the case of Ni and Fe, DFT calculations demonstrate that the molecules metalate at room temperature by incorporating a surface metal ad-atom when a monolayer is deposited on these substrates. The achievement of a complete metalation by Fe and Ni can be regarded as a test case for successful preparation of well-defined and controlled arrays of molecular-based magnets.

Finally, given the difficulty in evaporating some heavy metals (such us Rh, W, Ir or Ru) in UHV, a useful procedure to metalate the 2H-TPP in UHV has been shown by Papageorgiou et al. [18]. They demonstrated the formation of RuTPP on Ag(111) after the exposure of 2H-TPP to the carbonyl precursor molecules $Ru_3(CO)_{12}$ containing the metal and a thermal treatment to complete the reaction. This novel technique allows the engineering of new interfaces with geometrical arrangements of atoms for single-site catalysis or light-harvesting.

4. Conclusions

Surface coordination of organo-metallic systems onto metal substrates in UHV has attracted great attention in the recent years due to the prospective of applications like novel catalytic systems, light-harvesting antennas, sensors and molecular magnets. Many of these systems were notably explored using macrocyclic compounds, where the tetrapyrrole macrocycle may readily incorporate several metal ions.

Here we have described how the redox reactions needed to metalate the 2H-TPP monolayer on metallic surfaces can be performed, depending on the substrate and on the inserted metal ion, resulting in different systems with various characteristics. Metal evaporation, picking-up of a substrate atom and reaction with metal precursor molecules are the three methods that allow the creation of new metallo-porphyrins on metal substrates. Finally, since the physical and chemical properties of these metallo-organic systems depend on the coordinated metal ion but also on the interaction with the substrate, annealing treatments of the original 2H-TPP or the metalated TPP monolayer may change the conformation and the adsorption geometry of the molecule, therefore modifying both the physical and chemical properties of the system, even in the case of the same metallic center.

References

[1] Wöll C (ed) 2009 *Physical and Chemical Aspects of Organic Electronics* wiley ed

[2] Yokoyama T, Yokoyama S, Kamikado T, Okuno Y and Mashiko S 2001 *Nature* **413** 619–21

[3] Jung T, Schlittler R and Gimzewski J 1997 *Nature* **386** 696–8

[4] Teugels L G, Avila-Bront L G and Sibener S J 2011 *The Journal of Physical Chemistry C* **115** 2826–34

[5] Auwärter W, Weber-Bargioni A, Brink S, Riemann A, Schiffrin A, Ruben M and Barth J V 2007 *Chemphyschem: an European journal of chemical physics and physical chemistry* **8** 250–4

[6] Buchner F, Flechtner K, Bai Y, Zillner E, Kellner I, Steinrück H P, Marbach H and Gottfried J M 2008 *Journal of Physical Chemistry C* **112** 15458–65

[7] Kretschmann A, Walz M M, Flechtner K, Steinrück H P and Gottfried J M 2007 *Chemical communications* 568–70

[8] Gottfried J, Flechtner K, Kretschmann A, Lukasczyk T and Steinrück H 2006 *Journal of the American Chemical Society* **128** 5644–5

[9] Weber-Bargioni A, Reichert J, Seitsonen A, Auwarter W, Schiffrin A and Barth J 2008 *Journal of Physical Chemistry C* **112** 3453–5

[10] Shubina T E, Marbach H, Flechtner K, Kretschmann A, Jux N, Buchner F, Steinrück H P, Clark T and Gottfried J M 2007 *Journal of the American Chemical Society* **129** 9476–83

[11] Suto K, Yoshimoto S and Itaya K 2003 *Journal of the American Chemical Society* **125** 14976–7

[12] Haq S, Hanke F, Dyer M S, Persson M, Iavicoli P, Amabilino D B and Raval R 2011 *Journal of the American Chemical Society* **133** 12031–9

[13] Gottfried M and Marbach H 2009 *Zeitschrift für Physikalische Chemie* **223** 53–74

[14] Hanke F, Haq S, Raval R and Persson M 2011 *ACS nano* **5** 9093–103

[15] Lukasczyk T, Flechtner K, Merte L, Jux N, Maier F, Gottfried J and Steinruck H P 2007 *Journal of Physical Chemistry C* **111** 3090–8

[16] González-Moreno R, Sanchez-Sanchez C, Trelka M, Otero R, Cossaro A, Verdini A, Floreano L, Ruiz-Bermejo M, Garcia-Lekue A, Martin-Gago J A and Rogero C 2011 *The Journal of Physical Chemistry C* **115** 6849–54

[17] Nowakowski J, Wäckerlin C, Girovsky J, Siewert D, Jung T A and Ballav N 2013 *Chemical communications* **49** 2347–9

[18] Papageorgiou A C, Fischer S, Oh S C, Sağlam O, Reichert J, Wiengarten A, Seufert K, Vijayaraghavan S, Ecija D, Auwärter W, Allegretti F, Acres R G, Prince K C, Diller K, Klappenberger F and Barth J V 2013 *ACS nano* **7** 4520–6

[19] Yokoyama T, Yokoyama S, Kamikado T and Mashiko S 2001 *The Journal of Chemical Physics* **115** 3814

[20] Goldoni A, Pignedoli C A, Di Santo G, Castellarin-Cudia C, Magnano E, Bondino F, Verdini A and Passerone D 2012 *ACS nano* **6** 10800–7

[21] Fukuzumi S 2000 *Journal of Porphyrins and Phthalocyanines* **4** 398–400

[22] Diller K, Klappenberger F, Marschall M, Hermann K, Nefedov A, Wöll C and Barth J V 2012 *The Journal of chemical physics* **136** 014705

[23] Diller K, Klappenberger F, Allegretti F, Papageorgiou a C, Fischer S, Wiengarten A, Joshi S, Seufert K, Écija D, Auwärter W and Barth J V 2013 *The Journal of chemical physics* **138** 154710

[24] Di Santo G, Sfiligoj C, Castellarin-Cudia C, Verdini A, Cossaro A, Morgante A, Floreano L and Goldoni A *Chemistry - An European Journal* 12619–23

[25] Di Santo G, Castellarin-Cudia C, Fanetti M, Taleatu B, Borghetti P, Sangaletti L, Floreano L, Magnano E, Bondino F and Goldoni A 2011 *The Journal of Physical Chemistry C* **115** 4155–62

[26] Chen M, Feng X, Zhang L, Ju H, Xu Q, Zhu J, Gottfried J M, Ibrahim K, Qian H and Wang J 2010 *The Journal of Physical Chemistry C* **114** 9908–16

[27] Buchner F, Seufert K, Auwärter W, Heim D, Barth J V, Flechtner K, Gottfried J M, Steinrück H P and Marbach H 2009 *ACS nano* **3** 1789–94

[28] Buchner F, Schwald V, Comanici K, Steinrück H P and Marbach H 2007 *Chemphyschem : an European journal of chemical physics and physical chemistry* **8** 241–3

[29] Brede J, Linares M, Kuck S, Schwöbel J, Scarfato A, Chang S H, Hoffmann G, Wiesendanger R, Lensen R, Kouwer P H J, Hoogboom J, Rowan A E, Bröring M, Funk M, Stafström S, Zerbetto F and Lazzaroni R 2009 *Nanotechnology* **20** 275602

[30] Bai Y, Buchner F, Wendahl M, Kellner I, Bayer A, Steinruck H P, Marbach H and Gottfried J 2008 *Journal of Physical Chemistry C* **112** 6087–92

[31] Buchner F, Xiao J, Zillner E, Chen M, Röckert M, Ditze S, Stark M, Steinrück H P, Gottfried J M and Marbach H 2011 *The Journal of Physical Chemistry C* **115** 24172–7

[32] Borghetti P, Di Santo G, Castellarin-Cudia C, Fanetti M, Sangaletti L, Magnano E, Bondino F and Goldoni A 2013 *The Journal of Chemical Physics* **138** 144702

[33] Di Santo G, Blankenburg S, Castellarin-Cudia C, Fanetti M, Borghetti P, Sangaletti L, Floreano L, Verdini A, Magnano E, Bondino F, Pignedoli C A, Nguyen M T, Gaspari R, Passerone D and Goldoni A 2011 *Chemistry - An European Journal* **17** 14354–9

[34] Cudia-Castellarin C, Vilmercati P, Larciprete R, Cepek C, Zampieri G, Sangaletti L, Pagliara S, Verdini A, Cossaro A, Floreano L, Morgante A, Petaccia L, Lizzit S, Battocchio C, Polzonetti G and Goldoni A 2006 *Surface Science* **600** 4013–7

[35] Castellarin-Cudia C, Borghetti P, Di Santo G, Fanetti M, Larciprete R, Cepek C, Vilmercati P, Sangaletti L, Verdini A, Cossaro A, Floreano L, Morgante A and Goldoni A 2010 *Chemphyschem: an European journal of chemical physics and physical chemistry* **11** 2248–55

[36] Goldoni A *et al.* – *In preparation*

Journal of Physics: Conference Series **470** (2013) 012013 doi:10.1088/1742-6596/470/1/012013

Multiferroicity in V-doped $PbTiO_3$

F Ricci and V Fiorentini

CNR-IOM and Dept. of Physics, University of Cagliari, Cittadella Universitaria, 09042 Monserrato (CA), Italy

E-mail: frankyricci@gmail.com, vincenzo.fiorentini@gmail.com

Abstract. We report *ab initio* predictions on the proper multiferroic (ferromagnetic, insulating and ferroelectric) character of $PbTiO_3$ doped with vanadium (V). Ferromagnetically coupled V impurities carry a magnetization of 1 μ_B each. The coupling is expected to be strong, since the paramagnetic solution is higher by 150 meV/vanadium, and no stable antiferromagnetic solution was found. The electronic gap in the doped system is about 0.2-0.3 eV in the generalized gradient approximation (GGA), hence the system is properly multiferroic. V doping increases the spontaneous polarization in $PbTiO_3$, with an approximate percentual rate of 0.7 $\mu C/cm^2$.

1. Introduction

Modern technology uses extensively ferroic materials –i.e. having a spontaneous and permanent order parameter of one kind or another– with fairly exotic properties. For example, commodity data-storage devices exploit ferromagnets with giant magnetoresistance in magnetic random-access memories, in a continued push toward increased information density and reduced dimensions and writing energy. Besides their traditional use in sensors and actuators, ferroelectrics –with their permanent and switchable electrical polarization– are the building block of non-volatile, high speed, random-access memories which promise performances superior to semiconductor flash memories.

While their existence has been known since the 1960s, multiferroics, i.e. materials where polarization P and magnetization M coexist [1, 2], have enjoyed a renaissance in recent years, because of great improvements in growth techniques and new theoretical approaches. The coexistence of several order parameters and their mutual coupling may open the way to new device concepts, including the electrical addressing of magnetic memories without magnetic fields and their generating currents; the creation of multi-layer multi-state logical devices, exploiting combinations of polarization and magnetization; and magnetoelectric sensors. However, most of the current multiferroics lack sufficiently strong polarization or magnetization, or sufficient magnetoelectric coupling, or the correct kind of magnetic order (ferromagnets are desired, but antiferromagnets are much more common).

Some of the difficulties of single-phase multiferroics are bypassed by "metamultiferroics" built of multilayer heterostructures of ferromagnets and ferroelectrics [3], as for example in multiferroic tunnel junctions (MFTJ), where ferromagnetic electrodes are separated by a ferroelectric insulating barrier. The resistance against tunneling across the junction depends on the relative orientation of magnetization of the two electrodes (magnetoresistance) and on the direction of ferroelectric polarization of the insulating layer (electroresistance). This enables in principle a

4-level switchable resistance using external electric and magnetic fields. Of course, a further magnetic degree of freedom in the insulating-barrier material could potentially duplicate the number of states, and any chance of locally tuning the magnetization would add further design leeway. Following this line of thought, in this paper we report an *ab initio* investigation strongly suggesting that $PbTiO_3$ (PTO) doped with V (V:PTO) is a ferromagnetic small-gap insulator, hence a proper multiferroic. This is particularly interesting in view of the giant electroresistance effect predicted in tunnel junctions with PTO as ferroelectric tunnel layer [4].

2. V doping of PTO

Here we report the first principles prediction of a multiferroic state of lead titanate doped with magnetic vanadium. PTO is a well-known tetragonal perovskite with a high spontaneous polarization (86 $\mu C/cm^2$) of displacive origin. $PbVO_3$ also happens to be tetragonal, and with an even higher spontaneous polarization (152 $\mu C/cm^2$). Previous studies have shown its ground state to be an antiferromagnetic insulator (C-type), making it not especially interesting as a multiferroic. Motivated by our previous study of V-doped ferroelectric titanates [5], we examine the magnetic properties of vanadium diluted within the robust PTO ferroelectric.

Ab initio calculations are performed within density functional theory in the generalized gradient approximation (GGA) by Perdew-Wang using the PAW method [6, 7] as implemented in the VASP code [8]. Standard cutoff is used for the plane wave basis, and the k-point mesh for the bulk is 8×8×8 (appropriately rescaled for defect super cells). The Berry phase technique is used to calculate the polarization change upon V doping, using strings of 16 points in the polarization direction.

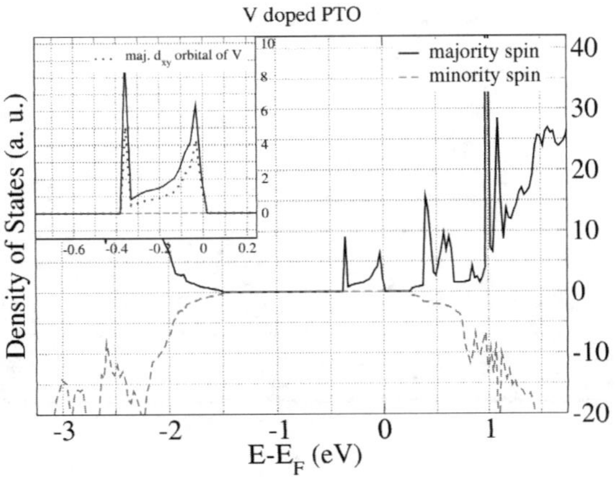

Figure 1. Density of states for V:PTO. Positive curve is majority spin, negative is minority spin. Fermi energy is at 0 eV. Inset shows a zoom of the gap peak. The dotted line is a V *d*-up.

After studying PTO and PVO in the respective ferroelectric phases, we substituted V for one Ti in a 2×2×2 tetragonal supercell of ferroelectric PTO, i.e. dopant concentration ~12.5%. A selection of structure parameters (lattice constants, distance of V or Ti from neighbouring O along the polar axis) are collected in table 1. V has one excess electron compared to Ti, so it is, as expected, spin-polarized with a moment of 1 μ_B. Orbital angular momentum is assumed to be quenched. Figure 1 displays the key result, i.e. the density of states (DOS) of ferromagnetic V:PTO. As we can observe, a fully occupied peak appears in the gap of PTO, placed so that a gap of about 0.2 eV survives. This DOS feature is obviously related to the excess V electron occupying antibonding conduction states. Indeed, as shown in the inset, the main contribution to that peak is that of spin-up *d* orbitals of V. Thus, PTO doped with vanadium is a small-gap

Table 1. Cell parameters for V doped PTO, bulk PTO and PVO. Theoretical values are calculated for this work in GGA approximation. Experimental values are from references [9, 10]

	V:PTO	PTO (Theor. / Expt.)	PVO (Theor. / Expt.)
a	7.848	3.924 / 3.895	3.806 / 3.804
c	8.351	4.175 / 4.171	4.979 / 4.677
V-O_{top}	1.709	-	1.700 / 1.677
V-O_{down}	2.377	-	3.278 / 3.01
Ti-O_{top}	1.792	1.795 / 1.75	-
Ti-O_{down}	2.380	2.380 / 2.42	-

ferromagnetic insulator thanks to the unpaired electron of dopant vanadium. The gap of PTO is of course underestimated due to the known gap error of semi-local functionals [12]; it may well be that the small V-related gap is also somewhat larger than calculated due to the same effects (self-interaction, xc discontinuity, etc.).

In the cell just discussed, V is ferromagnetic by construction, being coupled to its periodic images. To study an antiferromagnetic configuration we consider two V's at the same concentration in a $2\sqrt{2}\times2\sqrt{2}\times2$ tetragonal cell. The two V are placed in that cell as far as possible from each other. While the ferromagnetic configuration is stable, the antiferromagnetic configuration is not. The V moments disappear and the system converges to a metallic Pauli paramagnet of zero moment, which is about 300 meV higher in energy than the ferromagnet. Of course, this prevents the evaluation of a coupling parameter for magnetic models, but the ferromagnet seems quite stable nevertheless.

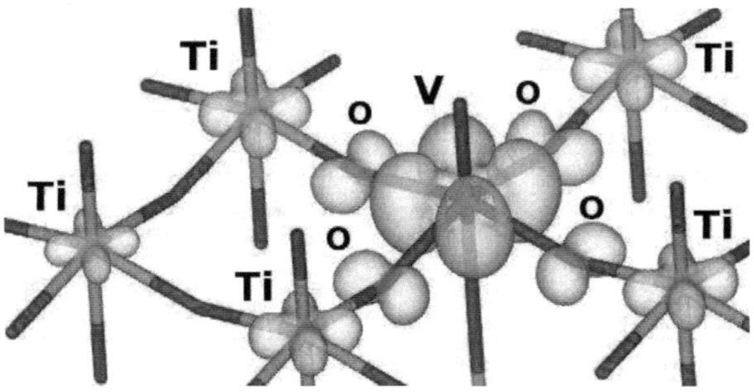

Figure 2. Majority spin density of the spin-polarized electron of V in the basal plane (iso-level: $10^{-3}\,e/\text{Å}^3$). The state is anti-bonding with first-neighbor O's, has d_{xy}-like nodal structure, and propagates through d_{xy}-like Ti states. Charge paths along the vertical axis are negligible.

Inspection of the charge density of the impurity state (Fig. 2) suggests a d_{xy}-like nodal structure and coupling of V's via V-O-Ti-O-... paths in the basal plane, whereas hardly any density is to be found along the vertical axis. While the V-centered state is, as expected, anti-bonding with first neighbors, the bonding O-V states (which mimic those of the substituted Ti) appear to make for a larger charge accumulation in the V-O bond region (Fig. 3), matching the shorter V-O bond length compared to Ti-O (1.71 vs 1.79 Å).

Following Ref.[11], the equilibrium concentration is $[\text{V}]=N_s \exp\left(-\text{E}_\text{f}\left(\text{V}\right)/\text{kT}_\text{g}\right)$, where N_s is the density of Ti sites and T_g the growth temperature. Assuming that the solubility limit for V in PTO is PVO, i.e. that high V contents will eventually cause phase separation of PVO, the formation energy is $\text{E}_f(\text{V})=\text{E(V:PTO)}-(n-1)\,\text{E(PTO)}-\text{E(PVO)}\approx1.23$ eV, involving the total

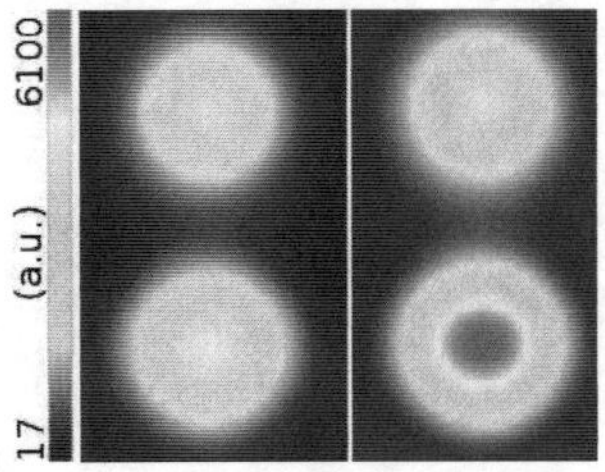

Figure 3. Total charge density (in arbitrary units) in the vertical plane across a Ti-O bond in PTO (left) and a V-O bond in V:PTO. A slightly larger bond charge can be appreciated in V:PTO.

energy of the doped supercell of V:PTO comprising n unit cells, and of the unit cells of PTO and PVO. Since N_s=1.25·10^{23}cm^{-3}, at a typical T_g=650 K we find a logarithmic concentration $\log_{10}$ [V] ≈13.4, which may increase perhaps to about 16 due to vibrational entropy (V is heavier than Ti). This thermodynamic solubility limit is relatively low, but it may well not apply whenever kinetics or non-equilibrium phenomena become dominant, as it happens in modern epitaxial growth techniques.

Finally, we estimated the changes in polarization induced by V doping. At 12.5% V we find a polarization increase of 8.3 μC/cm^2, which agrees nicely with 8.25 μC/cm^2 obtained by linear interpolation between the bulk values of PTO and PVO mentioned previously. Therefore, in terms of polarization, V does not seem detrimental, but in fact benign.

3. Conclusions

In summary, we presented an *ab initio* prediction of the proper multiferroicity of V-doped PTO. The gap is about 0.2-0.3 eV and the magnetization is 1 μ_B/V, or 1.5×10^{22} spins/cm^3 at our chosen V density. We predict the polarization to increase linearly with V concentration at the percentual rate of 0.7 μC/cm^2. Presumably this will only apply at relatively low concentrations, since the antiferromagnetic character of PVO will eventually take over.

Note – During review we have become aware of a very recent study [13] of transition-metal doping in a perovskite ferroelectric ($BaTiO_3$), reporting results in general agreement with the ones just presented.

References

[1] Wang K F, Liu J M and Ren Z F, *Advances in Physics* vol. **58**, No. 4, 321-448 (2009).
[2] Velev J P, Jaswal S S and Tsymbal E Y, *Phil. Trans. R. Soc.* A **369**, 3069-3097 (2011).
[3] Velev J P, Duan C-G, Burton J D, Smogunov A, Niranjan M K, Tosatti E, Jaswal S S and Tsymbal E Y, *Nano Lett.* **9**, 427 (2009)
[4] Ricci F, Filippetti A and Fiorentini V, *In preparation*
[5] Scarrozza M, Filippetti A and Fiorentini V, *Phys. Rev. Lett.* **109**, 217202 (2012).
[6] Blöchl P E, *Phys. Rev.* B **50**, 17953 (1994)
[7] Kresse G and Joubert D, *Phys. Rev.* B **59**, 1758 (1999).
[8] Kresse G and Furthmüller J, *Phys. Rev.* B **54**, 11169 (1996).
[9] Rodriguez J A, Etxeberria A, Gonzalez L and Maiti A, *J. Chem. Phys.* **117**, 2699 (2002).
[10] Ming X, Yin J-W, Wang X-L, Wang C-Z, Huang Z-F and Chen G, *Solid State Sci.* **12**, 938 (2010).
[11] Van de Walle C G, Laks D B, Neumark G F and Pantelides S T, *Phys. Rev.* B **47**, 9425 (1993).
[12] See e.g. Godby R W and Garcia-Gonzalez P, in *A Primer in Density Functional Theory*, Springer Lecture Notes in Physics; Fiolhais C, Nogueira F and Marques M A L eds. (Springer, Berlin 2003), Ch. 5, p.185.
[13] Chandra H K, Gupta K, Nandy A K and Mahadevan P, *Phys. Rev.* B **87**, 214110 (2013)

AUTHOR INDEX